A DIFFERENT RACE

A DIFFERENT RACE

*World War II, the Alaska Highway,
Racism and a Court Martial*

CHRISTINE & DENNIS McCLURE

LITTLE LANDS END PUBLISHING LLP • TAYLORS, SOUTH CAROLINA

Copyright © 2021 by Christine McClure and Dennis McClure

All rights reserved no part of this book may be reproduced in any form or by any electronic or mechanical means, including information storage and retrieval systems, without permission in writing from the copyright owner, except by a reviewer, who may quote brief passages in a review. Scanning, uploading and electronic distribution of this book or the facilitation of such without permission of the copyright owner is prohibited. Please purchase only authorized electronic editions, and do not participate in or encourage electronic piracy of copyrighted materials. Your support of the rights of the author, and other creative artists, is appreciated. Any member of educational institutions wishing to photocopy part of this work for classroom use, or anthology, should send inquiries to the publisher at the address set forth below.

ISBN: 978-1-7358417-0-0

Library of Congress Control number:

Published by Little Lands End Publishing LLP, Taylors, South Carolina

Final copy edit by Jill R. Hughes of The Editor's Mark, TheEditorsMark.Wordpress.com

Cover design by Tamian Wood, at www.BeyondDesignBooks.com

Interior page design and layout Stephen Tiano, Book Designer of Riverhead, NY, stiano@optonline.net

Printed in the United States of America

*To the black soldiers who served
with the 97th Engineering Regiment in Alaska*

CONTENTS

LIST OF MAPS		ix
ACKNOWLEDGMENTS		xi
INTRODUCTION		xiii
CHAPTER 1	Home	17
CHAPTER 2	The Army	25
CHAPTER 3	The War and the Plan	31
CHAPTER 4	To Alaska	39
CHAPTER 5	Getting Off the Ship	47
CHAPTER 6	Roadblocks	55
CHAPTER 7	Lytle and Green	61
CHAPTER 8	Disorganization and Getting Started	67
CHAPTER 9	Attack in the Aleutians	75
CHAPTER 10	Contractors at Valdez	79
CHAPTER 11	Mentasta Pass	85
CHAPTER 12	Change of Command	89
CHAPTER 13	Progress	95
CHAPTER 14	Reorganization	103
CHAPTER 15	Civilians in Uniform and the Oncoming 18th	111
CHAPTER 16	Beaver Creek	117
CHAPTER 17	A Miserable Winter	125
PHOTO GALLERY		133
CHAPTER 18	Black Troops Need Discipline	147
CHAPTER 19	Events at Big Gerstle River	157

CHAPTER 20	The Purpose of a Court-Martial	163
CHAPTER 21	Setting Up the Performance	169
CHAPTER 22	The Prosecution	175
CHAPTER 23	Adjusting the Story	185
CHAPTER 24	The Defense	193
CHAPTER 25	Summations and the Verdict	203
CHAPTER 26	Punishment	209
EPILOGUE	The Ten Back in the World	215
NOTES		221
SELECT BIBLIOGRAPHY		243

LIST OF MAPS

MAP 1	The Route of the Alaska Highway and Location of Seven Regiments	x
MAP 2	Alaska, Japan, and the Aleutian Islands	32
MAP 3	Richardson Highway and Abercrombie Trail Alaska	37
MAP 4	Valdez to Thompson Pass Alaska	50
MAP 5	Valdez to Slana Alaska	58
MAP 6	Slana to the Tanana River and Robinson's Crossing	69
MAP 7	The Old Alaska Highway White River To Big Delta	100
MAP 8	The Route of the 18th Engineers to White River Yukon, Canada	113
MAP 9	Winter Quarters	127

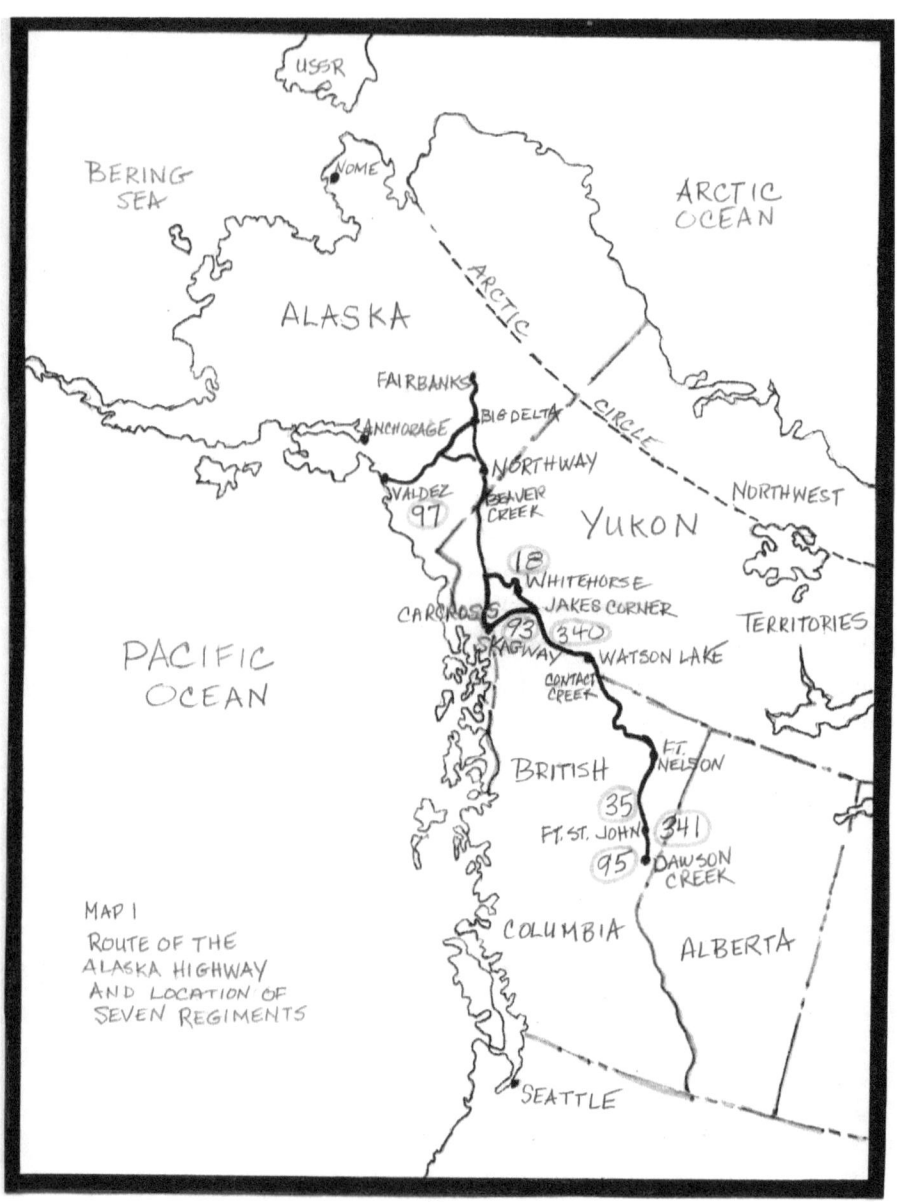

MAP 1   The Route of the Alaska Highway and Location of Seven Regiments

ACKNOWLEDGMENTS

Over the three years that we researched *A Different Race*, we accumulated a debt of gratitude to the many people who helped us find and understand the history we have written here.

In the summer of 2018, we were doing our best to find descendants of the men of the 97th Engineering Regiment and we found Walter Parsons III. We drove to Maryland's Eastern Shore to meet him and discovered that he had boxes full of documents that his father, the Captain Walter Parsons in our book, had saved from his time on the Alaska Highway project. His son let us load the boxes into our camper and gave us permission to copy and use anything we found in them for our book. We spent a week making copies, and this book would not have been possible without that information.

As we traveled in the Carolinas and Georgia in search of additional descendants and any information we could find about the men, we found helpful people and institutions everywhere.

In Norwood, North Carolina, we found the Norwood Branch Library, and that led us to Les Young, the expert on local history. Les worked hard to find people who might have information we would find useful and to set up appointments for us to meet and talk with them.

Billie Martin, Mayor Pro Tem of Marion, North Carolina, that summer helped us find and contact Fred Bryson, whose father served in the 97th.

And then we met Fred and his wife, Bonnie. They offered a treasure trove of memories, and Fred's father, Thad Bryson, appears several times in *A Different Race*.

In Alaska, as we traveled the route of the 97th from Valdez to the Tanana River, we met Carol Neeley, who remembered when the soldiers built the highway. Carol's father, Fred Neeley, owned the Chistochina Roadhouse located right on the route the soldiers took.

At the Camp Blanding Museum in Florida, George E. Cressman Jr. and his colleague Greg W. Parsons not only shared maps and information but actually gave us a guided tour of the portion of the camp where the segregated black soldiers lived and worked.

The staff at three libraries were particularly helpful—the Elberton, Georgia Library, the Wilson County Library in North Carolina, and the Salisbury Library in North Carolina.

In Elberton, Georgia, Jim Ree and his wife, Aurolyn Melba Hamm, run an African American History museum. They set up a meeting for us with several citizens of Elberton who might have remembered the families of Sgt. James Heard and Pvt. Willie B. Calhoun.

Pursuing the postwar career of Sims Bridges, we received invaluable assistance from the archives at Attica Prison in New York.

Finally, we visited the National Archives in Maryland and the National Archives Personnel Records Center in St. Louis, Missouri, several times. But there are always follow-up questions. In St. Louis, Lori Berdak Miller of Redbird Research LLC, and in Maryland, John Arnold PhD of Nicom Inc. helped us answer those.

INTRODUCTION

Early in 1942, in the wake of Pearl Harbor, America's leaders looked north with growing apprehension. The remote territory of Alaska and its Aleutian Island chain offered the Japanese a path to America, and America could not defend itself there. To get soldiers, weapons, and materials of war to Alaska, they needed a land route for convoys to cross northern Canada and enter the remote territory that in 1959 would become our forty-ninth state. The army dispatched seven regiments of the Corps of Engineers to build the Alaska Highway and gave them eight months to get it done.

From the railhead at Dawson Creek, British Columbia, the Alaska highway had to traverse sixteen hundred miles through the subarctic wilderness of British Columbia, Yukon Territory, and Alaska. Commanders divided the route into segments. Three regiments would work northwest from Dawson Creek. Two regiments would work southeast from Whitehorse, Yukon, to meet them. One regiment would work north from Whitehorse toward Alaska, and another regiment would work south through Alaska to meet them at the international border. The corps had only four regiments of white soldiers available, so three regiments of Black soldiers went north with them.

The officers who ran the army in those years, like many white Americans of the time, assumed that Black men lacked character, so Black soldiers served segregated in all-Black regiments commanded and tightly controlled by white officers. The Black soldiers who served in three segregated regiments on the Alaska Highway Project were of "A Different Race" from the white officers who commanded them. But the desperate rush to complete the highway by the end of 1942 was also a race—"A Different Race." Make no mistake about it, racism made the epically difficult race to complete the highway far more difficult than it had to be.

The seven regiments worked through the summer on their assigned segments, and in September the regiments working north from Dawson Creek met those working south from Yukon. In October one last gap remained and two regiments raced to close it. The white soldiers of the 18th Engineers clawed their way north out of Yukon Territory toward the Alaska border, and the Black soldiers of the 97th Engineers clawed their way south through Alaska to meet them there.

Snow covered the ground and kept falling. Temperatures ranged from zero to twenty degrees Fahrenheit. On October 25 the lead bulldozers of the two regiments met at Beaver Creek near the border, but the white soldiers of the 18th had bogged down miles to the south, struggling to build a road over permafrost. Col. Earl G. Paulis, northern sector commander, ordered the Black soldiers of the 97th to plunge on into Canada, building, through falling temperatures and accumulating snow, another fifty-five miles of road to the White River.

Determined to run convoys over the rough new road through the winter, senior commanders positioned soldiers along the way to keep it in repair and clear of snow and ice. They scattered the soldiers of the 97th at locations from the border all the way north to Big Delta. Unfortunately, no one had put a lot of thought or effort into providing housing at those locations.

On November 20, 1942, about 250 parka-clad soldiers, dignitaries, and newsmen stomped their feet and shivered through speeches at Soldier's Summit, celebrating the completion of the Alaska Highway. While the dignitaries shivered, the Black soldiers of the 97th moved to their winter locations. They would live in tents while they struggled to build their own barracks.

Natives and old-timers in Alaska and northern Canada remembered the winter of 1942–1943 as the worst since 1917. Temperatures during that winter reached seventy-two degrees below zero. Black soldiers huddled around stoves, turning like roasts on a spit to warm their cold sides. Frost covered interior walls and ceilings. The soldiers' threadbare uniforms offered little warmth. Some had only half of a sleeping bag. Some lacked parkas. They swapped and shared clothes, blankets, and sleeping bags, desperately trying to stay warm. They took turns going outside, cutting frozen tree trunks into firewood. They watched each other carefully, examining faces, noses, and earlobes for tell-tale signs of frostbite.

While the Black soldiers struggled to survive, the senior officers who commanded them worried. The Black soldiers of the 97th, they reasoned, like all Black soldiers, required especially strict discipline. In the rush to complete the road and then survive the winter, discipline had gone by the boards.

Commanders needed a virtuous example, and in March 1943 at a camp on the Big Gerstle River, ten young Black soldiers gave them an opportunity to set one. The temperature hovered around thirty-four degrees below zero when a young white officer ordered the ten to climb into the back of an unheated truck for a 130-mile trip. As the men eyeballed its unheated, ice-and-snow-packed cargo area with its ragged canvas cover, the cold that surrounded them took on ominous significance. Moving down the road would take the raw cold to a whole new level. And they had learned the hard way just how dangerous the Alaska cold could be.

To the young lieutenant's intense frustration, the soldiers feared riding in a dangerously frigid truck more than they feared him. He canceled the trip and ordered the men confined to quarters. Four days later, he preferred charges against all ten men for the most serious crime a soldier can commit—mutiny.

Officious young officers do these things. When they do, more senior officers take them aside and point out the error of their ways. A discipline problem isn't mutiny. If you need to punish the men's recalcitrance, have them peel potatoes or take an extra turn at guard duty.

Not this time. This time commanders needed a virtuous example for their Black soldiers. A formal investigation ensued, and in June a court-martial convened. One soldier, Willie Calhoun, had not even been present when the supposed offense occurred. The court-martial acquitted him. But the court-martial also ordered the army to dishonorably discharge nine soldiers and imprison them at hard labor: James Heard for twenty years; Sims Bridges for eighteen years; Lee Ratliff for twelve years; Willie Howell, and Robert Rucker for five years each; James Hollingsworth, Josh Weaver, Warren Lindsey, and Eugene Fulks for three years each.

On a frigid March morning at the Big Gerstle River, nine young men committed disobedience—clearly not a crime. On a June evening in Whitehorse, the United States Army perpetrated a gross miscarriage of justice.

To find the villain in this story we need to go back in time, back before the war changed everything.

CHAPTER ONE

Home

In the late 1930s, white and Black people shared cities and, especially, small towns across the American South. They grew up and learned about the world there. They met, married, and raised their children there. They earned a living there. They loved and hated, fought and reconciled, laughed and cried. But history had created in every one of those small towns not one but two towns, superimposed one atop the other. The citizens of the white town and the Black town shared a space and interacted with one another. But they didn't mingle. And the Black town existed in all but complete bondage to the white town.

> An invisible hand ruled . . . the lives of all the colored people in . . . the entire South. It wasn't one thing; it was everything. The hand had determined that white people were in charge and colored people were under them and had to obey them like a child in those days had to obey a parent, except there was no love between the two parties as there is between a parent and a child. Instead there was mostly fear and dependence—and hatred of that dependence—on both sides.[1]

Jim Crow, an elaborate system of laws and rules that absolutely segregated Black from white and absolutely established white dominance, prevailed in every Southern state. But Jim Crow went well beyond rules and laws. Granted absolute authority, whites could do whatever they liked to the Blacks among them. A white farmer paid his Black sharecropper whatever he liked and manipulated the contract between them as he chose.² A white man out on the town could drive through a Black neighborhood and demand of a young Black stranger that he find him "a nice clean colored girl."³ Whites could, with impunity, beat, shoot, or lynch Blacks for the slightest resistance to the most capricious demands, and Black parents raised their sons and daughters to stifle their resentment and submit.⁴ If a young man grew up resenting the system, he also grew up knowing he couldn't do anything about it.

The citizens of the Black towns had leaders. Across the South very few men and women managed to slip around the "invisible hand" and get themselves a bit more educated than most. The towns had businesses that served Black customers, preachers who preached to Black congregations, dentists, and even doctors. These Black citizens chafed at the unfairness of the system—chafed not only for themselves but also for their Black fellows. Sometimes they led the Black community to push back, at least a little bit. Invariably the white community saw to it that these Black leaders paid a price for their temerity.

James Heard grew up in Elberton, Georgia, where Black physician James Thompson pushed a little too hard and a little too often. First, Dr. Thompson inaugurated an annual Emancipation Day celebration where Black Elberton could gather, listen to speeches, and celebrate the end of slavery.⁵ The white citizens of Elberton did not approve. Dr. Thompson then went further by taking on the local landowners, who routinely bought life insurance policies on their sharecroppers and tenant farmers, naming themselves as the beneficiaries. Dr. Thompson fought the practice by refusing to sign required insurance documents on his patients. When, in 1915, Dr. A. S. Oliver, who was white, shot and killed Dr. Thompson, authorities at least arrested him. But no witnesses bothered to testify at his trial, which lasted exactly one day. The all-white jury exonerated Oliver.⁶

Like most of Elberton's Black citizens, James Heard's mother, Sally Mae Heard, had not been educated past the sixth grade. She did manage to find a husband and bring James Heard and eleven siblings into the world. James was born in 1923.⁷ Sally's husband, Chris Heard, labored intermittently for white farmers. He brought in a bit of money but spent most of it on booze. When Dad did his drinking in public, he got the attention of the police or sheriff. More than a few times his public drinking landed him in jail. Dad came and went in James's world. Sally anchored it. When Sally evicted her husband permanently, nine-year-old James's life didn't change much. White families had children to raise, and houses to clean, and laundry to do, and Sally did all of that for them, piecing together a meager living for herself and her twelve children. At home, of course, she cleaned and did laundry and tended to her own children. And for a long time, she had put up with her errant husband.

In 1930 James turned seven and Sally bundled him off to school—a Black school, of course, with a Black teacher. Nobody, including Sally, expected school to do much for James. He had to walk to and from the school—a long way. Some days he didn't make it. Other days a white farmer might offer work—a bit of income more important than a few days at school. James didn't find school all that inviting anyway.

Georgia in the 1930s offered Black children very little in the way of school. Students walked as many as seven miles. The state considered one room enough for a Black school. If a local church or empty cabin offered the room, the state saw no reason to spend money building a school for Blacks. Nor did the state feel it necessary to clean and maintain these schools. Black school yards across the state varied from untidy to positively filthy. The state budget determined the length of the school year, which for Black schools could be as short as four months. When the state bought books for white schools, they handed down the used ones to the Black schools. In 1940 the state spent $36.29 on each of its white students but only $4.59 on its Black ones.[8]

James went to school—sort of. He reached the third grade but failed twice. Then he failed fourth grade. He liked his teacher and got along with his fellow students, but he didn't learn a lot. He played pickup baseball and football. And he learned to gamble. Elberton's authorities frowned on gambling, and gambling got James arrested at sixteen.[9] He found a job along the way—making deliveries for a drugstore—but it didn't pay much. When he turned eighteen, he found a job in a laundry that paid eight dollars a week. Pulling the plug on school, he went to work full time. He kept three dollars a week for himself and gave the other five dollars to his mother for room and board.[10]

Occasionally he screwed up and got the attention of white authorities. In 1940 he got in a fistfight and spent a few hours in jail. That same year the police decided he had committed grand larceny. They arrested him, kept him in jail for a few hours, but then changed their minds. He got good at gambling, and he loved it. In 1941 police caught him at that again, kept him in jail overnight, and then fined and released him.[11]

As a young man, James had a place in the Black community. He played a little baseball, a little football. He worked at the laundry. He hung out with his friends and liked girls, dancing, and movies. Gambling got him into trouble but not serious trouble. Most of his friends gambled, and the authorities expected and tolerated it—that is, if it didn't affect the white community, and it rarely did.

The unfairness of life in Elberton didn't escape James Heard. White families had more. White people made the rules, and if you broke the rules they could hurt you. They could hurt you a lot. He tried to avoid white people.

Sixty-four miles separate James Heard's Elberton from Willie Howell's Madison, Georgia, and the history of the South cut both communities from the same cloth. Just eight years old when his father died, Willie lived with his grandmother, his mother, and two sisters. Like James's mother, Willie's eked out a precarious living cleaning houses, doing laundry, and tending the children of families in the white community. Willie's mother and grandmother attended church regularly and insisted that Willie and his sisters do the same.[12]

A refuge from the hostile white world, church played a special role in Black communities. Black members worshipped God together in their own way, free of white rules. They sang not hymns but spirituals, passed down from their slave ancestors. Black preachers sang their sermons, and emotional Black members punctuated them with fervent responses—"amen," "glory be," tears, waving arms, and moving bodies.[13]

In church, free of Jim Crow, members could safely express their thoughts and opinions. They could even strategize together to deal with the depredations of the white community. Most of all, church members had each other's backs. Those who had little shared with those who had nothing. They survived in a hostile and sometimes downright dangerous world, not as individuals but as a group.[14]

When Willie turned six, his mother duly sent him off to school, a Black school like James's school in Elberton. He worked part time but couldn't earn much. Turning sixteen, at the end of the eighth grade, he exchanged home and school for a Civilian Conservation Corps (CCC) camp in Monticello, Georgia. Ultimately, from Monticello, he moved on to a job with the Works Progress Administration (WPA).[15] The unfairness of life in Madison and, later, the CCC and the WPA didn't escape Willie, but he couldn't do much about it. He stuck with his own, avoided white folks if he could, and presented a carefully sculpted persona when he couldn't. Safety and security came from the Black community.

Willie Calhoun grew up in the same part of Georgia as James Heard and Willie Howell. At one point he lived just a couple of miles from Heard in Elberton.[16] Ben and Mary Calhoun had three kids. Willie had a younger sister and, for a while, a younger brother. Accidentally poisoned, Willie's little brother died very young. Ben Calhoun had a serious alcohol problem, and at age seven Willie went to live with his uncle and aunt, John and Lilla Taylor, on their farm in nearby Jeffersonville.[17]

Like Heard and Howell, Willie Calhoun did his time in school. As a teenager he worked part time as a drugstore clerk and then as a grocery delivery boy. He paid "too much attention to the girls."[18] And he frequently got into fights. But he played football and baseball. Having little interest in academics, he failed three grades and left school behind at seventeen. In 1940 Willie worked on road construction with the WPA in Macon County.

James Hollingsworth and Josh Weaver both grew up in rural, small-town Georgia in places very much like those that nurtured Heard, Howell, and Calhoun. James grew up with six siblings on a farm and completed grammar school but spent a lot more time working the farm. Josh also grew up on a farm, completed grammar school, and in 1940 worked in a sawmill.[19]

If Georgia's history had superimposed Black and white communities in its towns and cities, North Carolina's similar history had done the same in Lee Ratliff's Norwood, Warren Lindsey's Elm City, and Robert Rucker's Salisbury. Young Lee, Warren, and Robert resented the system but knew not to question or even think much about it.

Lee Ratliff's widowed mother, Hannah, lived in Waddletown, one of three Black neighborhoods in Norwood. She eked out a living working as a domestic servant for Otto and Elizabeth Mabry. A thin, wiry, and energetic Black lady, Hannah usually wore a white apron and, on special occasions, a white hat. She canned fruits, vegetables, and meats. She cleaned the house. She cooked the meals. She made soap from grease and lye and washed the Mabrys' clothes in a big iron kettle sitting over a fire in the backyard. She didn't have a lot of time for Lee. But she kept him fed and clothed.[20] North Carolina in the 1930s offered its Black children pretty much the same education Georgia offered. Lee made an attempt at school but left after the ninth grade.

Tobacco dominated Warren Lindsey's Wilson County.[21] And tobacco dominated his childhood. Growing and curing bright leaf tobacco involves miserable hard labor and lots of it. John and Carrie Lindsey produced tobacco, and their eight boys and five girls came to the fields as early as possible. When John died in 1932, Carrie exercised the only choice she had—she kept going. Warren quit school after the seventh grade.[22]

In 1930 ten-year-old Warren's Black community in Wilson County experienced the most terrifying of the rituals that white Southerners routinely visited on their Black neighbors. Oliver Moore, a Black tenant farmer, made a serious mistake. He asked his landlord for back wages, and shortly found himself in the Wilson County jail, charged with assaulting the landlord's daughters. That night approximately two hundred white men assembled outside the jail. Masked men came inside, took Moore from his cell, and drove him a little way out of town. "Once there, the group placed two plow lines under his arms, tied a rope around his chest, suspended him from a tree and shot him over two hundred times."[23]

Sims Bridges came of age in Pritchard, Alabama.[24] Eugene Fulks grew up in a larger town—Vicksburg, Mississippi.[25] The two men grew up with the same invisible hand. White men in Mississippi, in fact, took the invisible hand to an extra level of arbitrary viciousness. Black men were lynched in every Southern state, but white Mississippians turned to that "solution" more passionately and more frequently.[26]

Robert Rucker's family moved regularly. But they moved from one small Southern community to another and never left the invisible hand behind. Robert left school after the fourth grade.[27]

In 1939 young men across the South, individuals all, nevertheless shared an implacable reality. "From the early years of the twentieth century . . . , nearly every Black family in the American South, which meant nearly every Black family in America . . . were all stuck in a caste system as hard and unyielding as the red Georgia clay."[28]

CHAPTER TWO
The Army

In 1940 the United States slowly, reluctantly prepared to go to war, and young men all over the country stood on a precipice. Congress passed, and President Franklin Roosevelt signed, a draft law that required every young man in America to register. James Heard, the nine men who would make up his squad, and a little over a thousand more young Black men with similar backgrounds would go over that precipice and end up in the 97th Engineering Regiment—and in Alaska.

Willie Howell entered the army at Fort Benning, Georgia, on January 25, 1941. Eugene Fulks entered at Camp Shelby, Mississippi, in March. Warren Lindsey and Sims Bridges came to the army in April—Warren at Fort Bragg, North Carolina, and Sims at Fort McClellan, Alabama. James Hollingsworth, Josh Weaver, James Heard, and Willie Calhoun came to Fort Bragg in June. Lee Ratliff and Robert Rucker came to Fort Bragg in August.[29]

Throughout 1941, stunned, nervous young men came to reception centers all over the country. Sergeants bellowed at them, threw uniforms and boots at them, and made the point that they no longer belonged to their mothers, or their communities, or even themselves. They now belonged to the army. The army used its reception centers to sort out its new recruits, to group and count them, and then send the groups out to training centers, sometimes on the same post, sometimes at the other end of the country.

The army decided to make James Hollingsworth, James Heard, and Willie Howell engineers, loaded their groups into railroad cars, and took them up to the Engineer Training Center at Fort Belvoir, Virginia. In the course of three months, the segregated 8th Engineering Battalion introduced young Black soldiers to military courtesy, drill, weapons—and the rudiments of army engineering. The army also introduced them to marching, running, calisthenics, and obstacle courses. Not all the young men destined for the 97th Engineers and Alaska trained with the 8th, but most did.[30]

As the army trained the flood of new recruits, it also created the military units that would use them. In June 1941 at Camp Blanding, Florida, the seventeen white officers who would run the brand-new 97th Engineering Battalion got themselves organized and waited for their allotment of enlisted troops.[31] One of those officers, Cpt. Walter H. Parsons, grew up in Palestine, Texas, and joined the Reserve Officers' Training Corps (ROTC), during his college years at Texas A&M. He was commissioned in the Army Reserve when he graduated in 1930. He started his career, and his Army Reserve service receded to the background of his life. After holding a series of jobs through the early years of the Depression, he settled with Humble Oil and Refinery Company in 1933.[32] Along the way Walter met pretty Abbie Dawsey and they married in 1932. Walter's career kept the young couple moving—Galveston in 1932; Eagle Lake in 1933; Corpus Christi in 1934; Lake Charles, Louisiana, in 1936; Baytown, Texas, and then Ingleside in 1938. The couple had five children: Walter, David, Donald, Anne, and, finally, Michael. By 1940 Walter and his family lived on the edge of the same precipice as every other young man and young family in America. In April 1941 they went over it when the army called 1st Lt. Walter Parsons to active duty. Lieutenant Parsons went to Fort Belvoir, graduated from the 5th Instructor Course in June, and then moved on to Camp Blanding, in Florida.

Camp Blanding had mushroomed out of the north Florida sand and woods literally overnight in 1941. Construction workers and soldiers swarmed the post and the nearby town of Starke. In a matter of months, the army built ten thousand buildings and 125 miles of paved roads and installed 81 miles of water lines and over 250 miles of electrical wiring. Thousands of soldiers descended on Blanding to live and train in the middle of a vast construction site.[33]

Mostly Camp Blanding trained white soldiers. The 31st Infantry Division came in December 1940, and the 43rd Infantry Division came in March 1941. Together that was roughly forty thousand soldiers. But the army brought smaller units as well, some of them segregated Black units, including the 97th Engineering Battalion.[34] And race complicated things. The army didn't invent racism and segregation, but they did very little to correct those wrongs.

The *Pittsburgh Courier*, in its Saturday, July 19, 1941, edition, headlined an article about Camp Blanding "Whites Trained as Combat Soldiers; Negroes Do Work." The article noted that "most of the whites are organized into large units which have some military integrity. . . . Hundreds of negroes are being split up into service detachments or quartermaster detachments." The article offered hope for improvement with the imminent placement of a "regiment of colored soldiers" at Blanding. "But it must not be forgotten that the term 'engineers' in the Army may or may not be a euphemism, used to cover up something dirty."

The army didn't provide housing for the officers of the 97th, so Lieutenant Parsons found temporary quarters for his family at Blanding near Keystone Lake. The children loved it; they even learned to swim in the lake. In September the family moved to Jacksonville, forty miles away, and Parsons commuted. Enlisted men didn't bring their families to Blanding, of course, so the enlisted men of the 97th lived in military quarters. They lived in what the *Pittsburgh Courier* called "hutments," large pyramid tents with four-foot-tall framed walls topped with canvas. Along with the hutments, the army provided dayrooms, mess halls, an infirmary, and even recreation buildings.[35]

The army took the young Black men of the 97th out of their small Southern towns and introduced them to a wider world. They learned military discipline, trained hard, ate relatively well, saw movies, and developed friendships.[36] But segregation and basic unfairness came to the army with them. They slept in separate barracks, ate in separate mess halls, and watched their movies in theaters separate from those of their white counterparts. White officers commanded them, and white men made the decisions.

The men who came to the 97th, both officers and enlisted, had trained as individuals. At Blanding they trained as a unit, learning what to expect from life in the regiment and what the regiment expected of them. White officers who understood their jobs in theory learned to do them in practice. They began to learn what to expect from their colleagues—and, more important, from their troops.[37]

Among the young Black soldiers, some stood out in training. The army made them noncommissioned officers (NCOs), the military equivalent of civilian foremen. James Heard became Cpl. James Heard. The army trained NCOs to supervise their fellows at work—like civilian foremen. But the army went further. In the army a good NCO took care of his men both on and off the job. He applied discipline, of course, but he also had their backs. At Blanding the enlisted men of the 97th shared barracks, ate together, worked together, played together. They formed personal bonds, made close personal friends—and occasionally enemies. They teased each other, they occasionally fought with each other, but they depended on each other.

In November the army found useful employment for the 97th—at Eglin Army Airfield.[38] Still part of the army in 1941, the rapidly expanding Air Corps needed training facilities and located the biggest at Eglin Field in western Florida. Air Corps pilots would come to Eglin to test new aircraft and to practice gunnery and bombing. In 1941 Eglin mushroomed as frantically as Camp Blanding, and in November the 97th came, not to train but to build.[39] Their orders were to build seven large machine gun ranges, each several hundred feet wide and over a mile long. The soldiers finally experienced real construction—that is, until December 7. After Pearl Harbor the only soldiers on post with rifles, the men of the 97th, pulled guard duty instead of building ranges.[40]

Tech 5 Thad Bryson served in Company B of the 97th as a cook. He came to the army from Old Fort, North Carolina, at Fort Bragg in August

1941; moved to the 97th at Blanding; and served with the regiment through Eglin and Alaska. In 1943, after the Alaska Highway was completed, Thad's kitchen stove blew up in his face, seriously burning him and threatening his eyesight.[41]

Most white officers brought prejudice to the regiment but not racism. For the best of them, working with Black soldiers through 1942, solving problems together, and sharing the misery of living in Alaska blew prejudice away. After Thad's accident, the commander of Company B, Cpt. James Forrestal, visited him in the hospital tent. Thad asked to stay with the regiment, but Forrestal would have none of it. He insisted on sending Thad back to Walter Reed General Hospital for treatment that probably saved his sight. Forrestal also arranged for a medical discharge that sent Thad home to Old Fort.

Thad had grown up in a two-room house on a small farm. Nobody expected much from a young Black man, and Thad didn't expect much from himself. Like most of his fellows, he had done his time in the poor-quality schools the South offered a Black child; he finished Hudgins High School, but Hudgins didn't offer grades beyond nine. An unpainted, two-story frame building, it also didn't offer enough space for home economics classes. Students learned to cook on a potbelly stove in the yard.

In early 1942 Thad found himself cooking in a mess hall at Eglin Field. The Tuskegee Airmen had come to train at Eglin, and the pilots ate in the mess hall. These Black officers who had college degrees and talked "educated" flew airplanes! One day a young white lieutenant passed Black Maj. Benjamin O. Davis Jr., commander of the airmen, and ignored him by failing to render the required hand salute. Thad watched, open-mouthed, while the Black major chewed the white lieutenant out—and collected the required gesture of respect. Thad came to rigid attention and rendered his own salute. The major returned it, smiled ever so slightly, and winked.

Thad came home from Alaska and Walter Reed, married, and had children. Near the end of his life, his son asked Thad if he could live his life over again, what might he change. The old man grinned and said, "I might find me a way to get into college and maybe learn to fly airplanes."

CHAPTER THREE

The War and the Plan

In the late 1930s, the Roosevelt administration had turned its attention to looming military threats, and army and navy planners worried about the territory of Alaska. The string of islands known as the Aleutians curves like a slender sword, south and then west from Alaska eleven hundred miles into the North Pacific—marking the boundary between the North Pacific and the Bering Sea. Adak, Kiska, and Attu, the islands at the point of the sword, lie perilously close to Japan. The strategic importance of Alaska posed two enormous problems for the United States. First, the outpost had fifteen thousand miles of undefended coastline. Second, no transport route, other than the sea, existed between Alaska and the contiguous United States. They couldn't get enough men, equipment, and supplies to Alaska to defend it.[42]

On December 7, 1941, the Japanese bombed Pearl Harbor and followed up with a coordinated assault across the Pacific.[43] In Washington, DC, senior officials' strategic worries about Alaska turned to panic. The navy had suffered grievous losses, and a flood of new commitments stretched its resources to the breaking point. Clearly the navy could not guard the sea-lanes leading to Alaska and the Aleutians and at the same time ferry supplies to land-based defenders there.

America desperately needed a land route to its remote territory, and the issue made it to the top of the national agenda when Secretary of the Interior Harold Ickes raised it in a meeting of President Franklin Delano

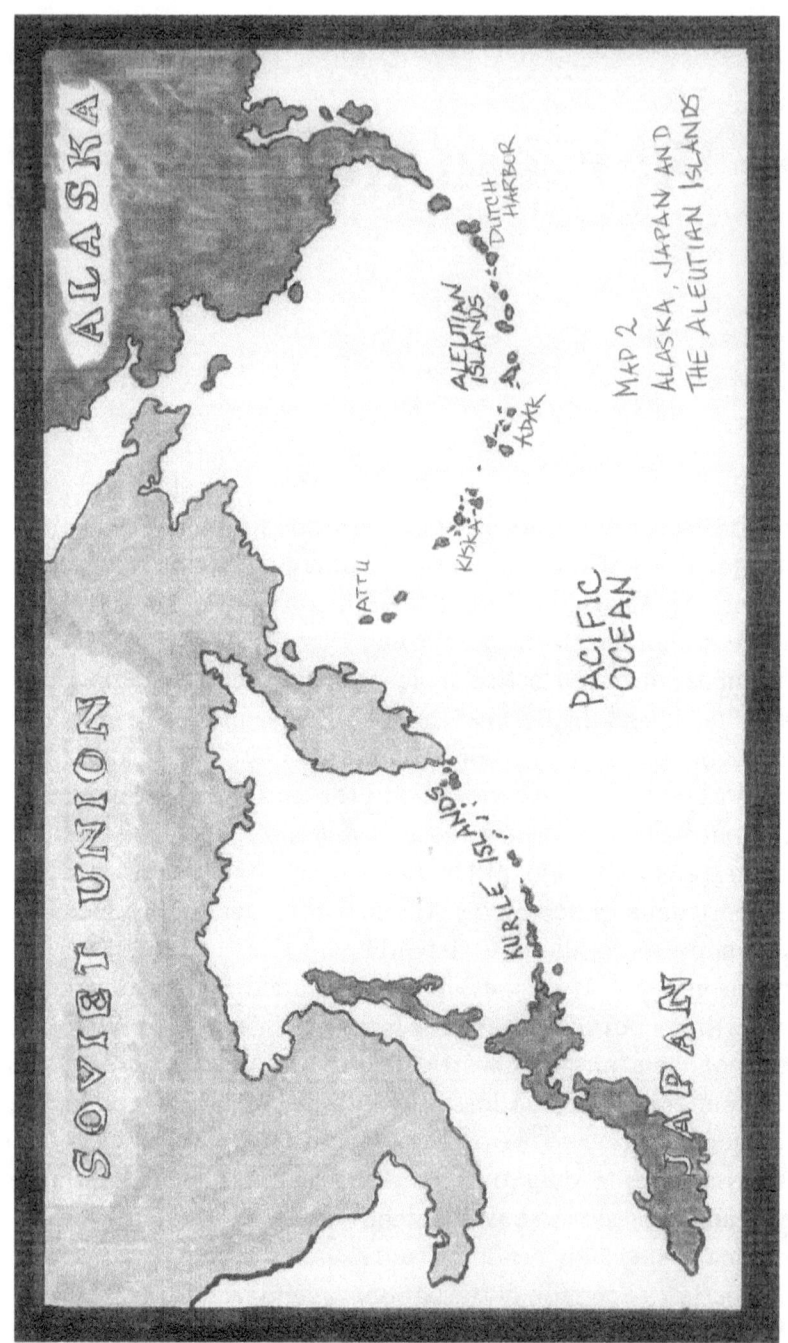

MAP 2   Alaska, Japan, and the Aleutian Islands

Roosevelt's cabinet on January 16, 1942.[44] The president directed the secretaries of War and Interior to investigate and find a solution. On February 2, 1942, Secretary of War Stimson met with Ickes and the secretary of the navy in his office to follow up on the president's instructions. Efforts on the ground in Canada and Alaska, they agreed, had to be under way before the spring thaw made it impossible to move in men and equipment.[45] The problem landed on the desk of Gen. Clarence L. Sturdevant at the Army Corps of Engineers, who on February 12 dispatched Col. William Hoge to conduct a reconnaissance on the ground in Canada and figure out the logistics of building the road.[46] Two days later, on February 14, the War Department directed the chief of engineers to begin construction of the highway at once.[47]

Up north, looking at the daunting challenge, Hoge realized immediately that he needed more than the four regiments Sturdevant had planned to send. Short of white soldiers, the corps would have to send segregated Black regiments north to the project. When Sturdevant wrote to inform Gen. Simon Bolivar Buckner, commander of United States Armed Forces in Alaska, that Black regiments would be working on the road, Buckner reluctantly accepted reality. He also demanded that Black men be kept away from the local population in Canada and Alaska.[48]

Experienced, thoroughly competent officers, Sturdevant and Hoge knew how to plan and execute an engineering project. But racism thoroughly contaminated not only their army but their thinking as well. They planned around engineering considerations; they also planned around racism and racist assumptions. Neither Hoge nor Sturdevant expected much more than "pick and shovel" work from what they assumed were incompetent Black men. Their plan had to make sure that competent white soldiers took the lead. Racism turned a relatively straightforward engineering plan into something very different.[49]

By March 3 Hoge had completed the reconnaissance mission. The project, changing on the fly, was under way. The mission required that the proposed highway, to be known as the Alaska-Canada Highway, or Alcan, follow the most direct route possible from the railhead at Dawson Creek,

British Columbia, to Fairbanks, Alaska. Hoge's plan divided the daunting sixteen-hundred-mile stretch into sections. He would bring the experienced and equipped 35th Combat Engineering Regiment through the railhead at Dawson Creek to build over the Canadian Rockies toward Yukon. The white 341st Engineers would follow them. Ultimately, Sturdevant would dispatch the Black 95th Engineers to follow the white regiments north, cleaning up and doing pick and shovel work.[50] The white 340th, the white 18th, and the Black 93rd would come in through Skagway, Alaska, and up into Yukon on the tiny, narrow-gauge White Pass and Yukon Route Railroad. The 340th would work south through Yukon toward the oncoming 35th. The 18th would work north through Yukon toward Alaska. The Black 93rd would follow the 340th.[51]

With the Black soldiers safely relegated to less demanding work and with plans in place to keep them away from the local population, General Sturdevant and newly promoted General Hoge turned their attention to the vexing problem of Alaska. If racist requirements and assumptions contaminated the plan for the road through British Columbia and Yukon, they also rendered the generals' plan for Alaska nothing short of ridiculous. Disaster loomed for the oblivious young Black soldiers and their equally oblivious white officers at Eglin Field.

The generals had only the Black 97th to go to Alaska and work south toward Yukon. They had no competent white regiment for the Black soldiers to follow with their picks and shovels. And they had no white regiment to work between the Black soldiers and the local population. The generals determined to solve their problem by surrounding the 97th with civilian contractors. Competent civilians would supplement the picks and shovels of the Black soldiers. In Alaska's existing, if primitive, road network, the generals thought they saw routing options that would let them keep civilian contractors between the Black men of the 97th and the local population. At some point it might have occurred to the generals to dispense with the 97th altogether. It didn't. It would take time to work through the Public Roads Administration (PRA) and set up the network of contractors, and the generals simply didn't have time. Sturdevant could expand the 97th from a battalion to a regiment and dispatch them to

Alaska very quickly. He set those wheels in motion and added the Alaska network of contractors to his requirements for the PRA in March, even as the generals devised a route for the Black soldiers and the civilians.

The doorway to northern Alaska had always been the Port of Valdez, and the United States Army had been looking for a better route from the doorway to Alaska's vast interior since the late 1800s. When, at the turn of the century, the Klondike Gold Rush turned the trickle of travelers through Valdez into a flood, the War Department had sent Cpt. William R. Abercrombie to create a better trail.[52] Travelers through Valdez came by ship up the inside passage, disembarked, and assembled on mudflats at the edge of the water. Beyond the mudflats, the land, folded by time and erosion, climbed abruptly to two enormous peaks separated by the snow and ice of the Valdez Glacier. Travelers loaded their supplies onto seven-foot-long sleds, dragged them across the mudflats, and lifted them onto the lip of the glacier with block and tackle. As they moved across the ice, they found deep crevasses; some travelers found them the hard way, plunging hundreds of feet to their death. In spring white fog shrouded the glacier, cracking ice boomed, and water poured off the face, sometimes trapping travelers for days at a time. Hands and feet froze. No one had enough to eat. Scurvy ran rampant.[53] From Valdez, Abercrombie by-passed the glacier. His trail climbed into the towering mountains by way of Keystone Canyon and Thompson Pass. From the pass his trail wound 350 miles through mountains and across rivers to Eagle, Alaska, on the Yukon River.[54]

The gold rush ended as quickly as it had begun, and when the army sent Maj. Wilds Richardson in 1905 to build an actual road, Fairbanks had replaced Eagle as the interior terminus. Richardson upgraded Abercrombie's trail as far as Gulkana and then turned northwest to the Tanana River and Big Delta and on to Fairbanks. At Gulkana he upgraded a branch from his main road to Slana. But from Slana northeast he left Abercrombie's trail untouched.[55]

Sturdevant and Hoge saw immediately that the Alaska Highway north out of Yukon, headed as directly as possible for Fairbanks, would intersect Richardson's road at Big Delta. From there to Fairbanks Richardson's road would do just fine. They needed only to build their Alaska Highway south from Big Delta to the Canadian border to meet the 18th Engineering Regiment working north through Yukon.

The Black soldiers in Alaska, of course, would have to come in country at Valdez, 270 miles from Big Delta.[56] The existing Richardson Highway offered relatively easy passage, but civilians in northern Alaska lived all along and in close proximity to the Richardson, and the generals had to keep the Black soldiers away from them. Looking at their maps, the generals found a solution. The Black soldiers would travel the Richardson Highway from Valdez to Gulkana, where it parted company with the old Abercrombie trail. From Gulkana they would continue northeast to Slana on state-maintained roads. From there, safely distant from local civilians, they would use the old trail, upgrading it just enough to get themselves and their equipment north to the Tanana River.[57]

White contractors would come in through Valdez and move north to Big Delta. From there they would build the highway south away from the civilians to a meeting with the Black soldiers on the north bank of the Tanana. Other white contractors would come in through Valdez and move north to Gulkana and Slana, prepared to follow the Black soldiers all the way north over Mentasta Pass to the Tanana, reinforcing the expected substandard efforts of the Black soldiers.

On March 31, 1942, a month before the 97th landed in Valdez, General Sturdevant penned a letter to Thomas McDonald of the PRA summing up the assistance the army required from the PRA and its civilians. The letter makes very clear that Sturdevant expected that the civilians would build the road and that the soldiers would play only a minor role:

My Dear Mr. McDonald:
    You are authorized to proceed as soon as practicable with actual construction of sections of the Canadian-Alaskan Military Highway described below . . .
        a. Widening, improvement or relocation of the existing road . . . from the Richardson Highway near Gulkana to the vicinity of Slana, Alaska . . .

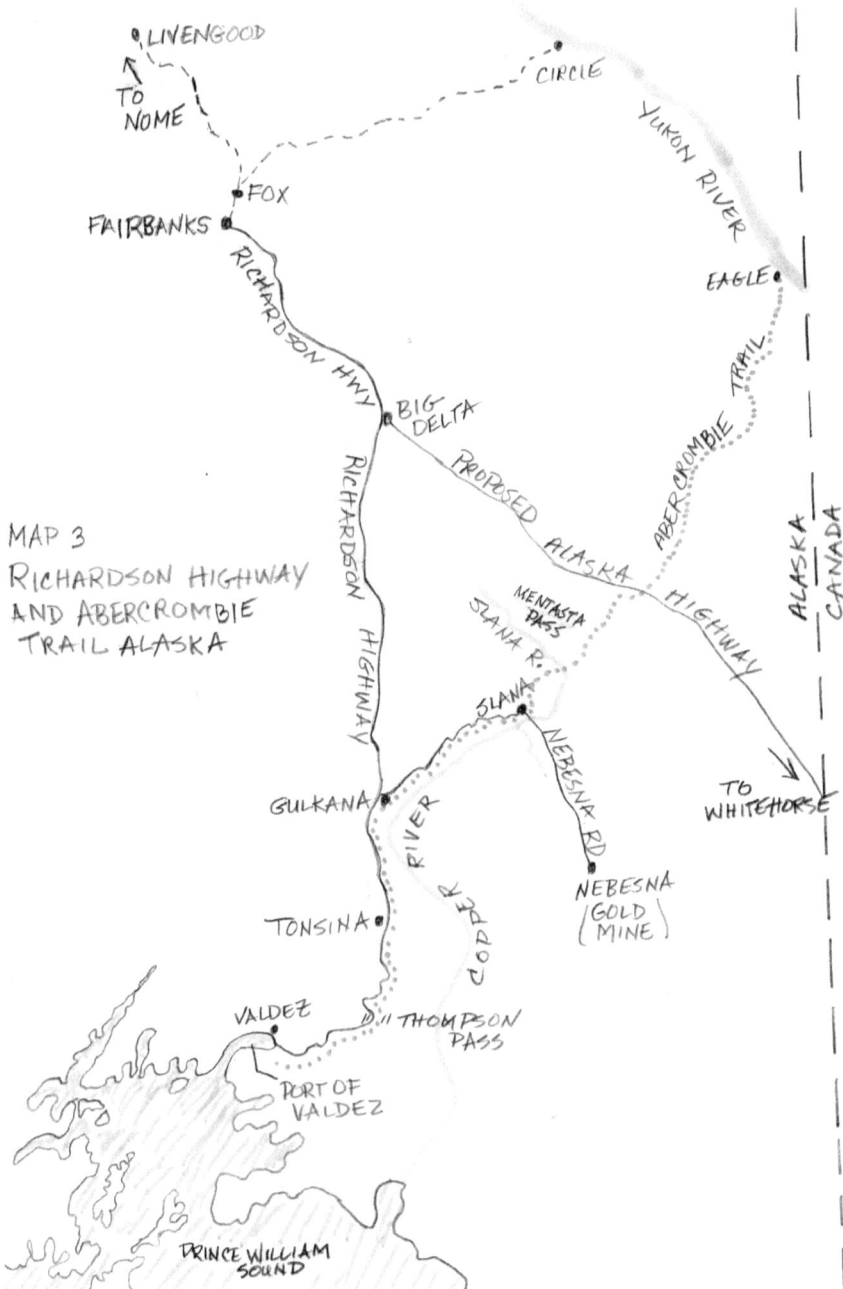

MAP 3 Richardson Highway and Abercrombie Trail Alaska

  b. Construction of a road in continuation of *a* above through Mentasta Pass and generally along the Tok River to a crossing of the Tanana River to be selected . . . *Engineer troops will construct a pioneer truck road throughout this section.*

  c. Construction of a road from the vicinity of Big Delta to a junction with *b* above . . . *Engineer troops will not operate in this section* . . .[59]

  The generals ordered the 97th to come in country at Valdez and move as quickly as possible up the Richardson Highway to Gulkana. At Gulkana they would leave the Richardson and follow the branch road to Slana. From Slana they would use Abercrombie's trail to get northeast to the Tanana River. They would have to upgrade the trail from Slana to the Tanana as they moved along, of course, but only just enough to accommodate themselves and their equipment.

  That upgrade from Slana to the river looked a lot easier on the generals' maps than it turned out to be on the ground, and it fell to the soldiers of the 97th to carry the plan through Valdez and into the harsh reality of Alaska. Luckily the Black soldiers delivered a lot more than the generals expected.

CHAPTER FOUR
To Alaska

Sturdevant and Hoge's plan, developing on the fly, reverberated through the Corps of Engineers. On March 9, 1942, orders from on high transformed the 97th Engineering Battalion at Eglin Field into the 97th Engineering Regiment, effectively doubling its size.[60] The new regiment struggled to incorporate new companies, hundreds of additional soldiers, and a shifting and expanding roster of officers.

Even the commanders of the 97th didn't learn their fate until April 7:

Secret
Immediate Action
War Department
The Adjutant General's Office
Washington

April 7, 1942

SUBJECT: Movement Order, Shipment 1864
TO: The Commanding Generals
Third Army; Western Defense Command
The Commanding Officer, Seattle Port of Embarkation
The Chief of Transportation Division, Services of Supply
The Chiefs of Supply Services

> It is desired that . . . take the necessary action without delay to prepare for foreign service and to move the units and detachments listed below to the Seattle Port of Embarkation. . . .
>
>   97th Engineers   1864-A   Camp Blanding, Fla   Off-46, WO-1, EM-1259
>   (Temporarily at Eglin Field, Florida) . . .
> * This unit has Negro enlisted personnel.
>
> DWIGHT D. EISENHOWER,
Assistant Chief of Staff,
Operations Division, W.D.G.S.

Col. Stephen Whipple took command of the 97th just in time to lead them out of Florida. Whipple had enlisted and served in the army during World War I. After the war he had enrolled at the University of California at Berkeley and in the ROTC. Upon graduating he took a job as a civil engineer for the California Water Commission and continued to serve in the Army Reserve. By 1934 he had risen to captain and by 1940 to major.[61]

Whipple learned management as a bureaucrat in the California Water Commission, and he brought a unique mind-set to his new command. As the battalion expanded to a regiment and prepared itself to leave Eglin Field for Alaska, Whipple hurriedly shuffled officers and assignments. For Parsons, now a captain, he created a brand-new position with the title of liaison officer. Busy subordinates chafed at endless staff meetings and Colonel Whipple's tendency to focus on details instead of on the larger picture.

The army promised that personal equipment for Whipple's men, including canvas cots, wool blankets, sleeping bags, clothing, cold weather gear, and other necessities, would meet them in Seattle. The army took most of the regiment's worn and outdated trucks away, promising that shiny new trucks would come to them through Seattle. They also promised that heavy equipment, including twenty D8 Caterpillar tractors, along with fuel and lubricants would follow them through Seattle and up to Alaska.[62]

Meanwhile, Walter and Abbie Parsons packed up their children and the family headed back to Texas. Walter packed his trunk, and twelve hundred Black enlisted men packed their barracks bags. The 97th convoyed to Pensacola, boarded two trains, and headed out of Florida on April 15.[63] They made it to Memphis, Tennessee, and St. Louis on April 16 and then to Kansas City and Dodge City on the 17th. A Hollywood movie called *Dodge City*, released in 1939 had made that city famous as the epicenter of the historic Wild West. "Since some of the boys had seen the picture show about the place, they got a big kick out of being there."[64]

As they arrived on the 18th in Leadville, Colorado, it snowed. "They opened all the windows they could and let the snow blow in. . . . They had snowball fights. It was the first time some of these boys had seen snow."[65] That same afternoon they got off the train in a Utah desert, got caught in a sandstorm, and had to get back on in a hurry. On the 19th they moved on to Idaho and Oregon, and they made it to Fort Lewis, Washington, on April 20, 1942.

Plans and movement orders look very good on paper, but reality at the Port of Seattle looked very different. Wartime logistical chaos prevailed, and the promised personal equipment for Whipple's men, including sleeping bags, winter clothing, and cold weather gear, was nowhere in evidence.[66] Whipple and his regiment, up against a brutal schedule, couldn't wait around. Whipple left Captain Parsons at the port to receive equipment and supplies and then ship them on up the inside passage to Valdez. The colonel and the rest of his regiment would sail north to get organized and wait for the shipments.

One company moved directly from Fort Lewis to the Seattle Port of Embarkation and moved the few small trucks the army had allowed them to bring along off the trains and onto the troopship USS *David Branch*, where they were chained to the deck. Through the day on April 22, the rest of the regiment moved to the port and crammed their twelve hundred bodies and barracks bags into the fetid hold. "Each man followed the man in front of him through a maze of hatches and companionways."[67]

The *David Branch* left port on the evening of April 22 and headed out into Puget Sound. From the Sound, it made its way north into Canadian waters along Vancouver Island and mainland British Columbia, then through the North Pacific, the Gulf of Alaska, and Prince William Sound.

Officers who got to go up on deck enjoyed spectacular scenery.[68] A few of the enlisted men aboard were also able to go on deck during the trip, but most of them didn't, remaining confined below in "a forest of steel pipes supporting canvas strips stretched tightly with ropes." The canvas strips, serving as hammocks, were "tiered three high." The face of the man on top grazed "a tangle of pipes. . . . The men below had to contend with the indentation made by the bodies of the men above."[69] They ate twice a day, lining up for hours to get food. "Food was dumped onto the mess tray and you proceeded to a chest-high table running the width of the ship. Once there, you moved along the table, eating as you went."[70]

The ship left Seattle in spring. A week later, sixteen hundred nautical miles north, approaching the Arctic Circle, winter still gripped Valdez, Alaska. Young men from the Carolinas and Georgia were poised on the edge of a very different world. Valdez existed as the port of entry for a transportation system that served the rugged northern interior of Alaska (see Photo 1). In other words, Valdez connected the oceans of the world to the Richardson Highway. The Valdez Dock Company, owned and managed by Robert Kelsey, provided a long wooden dock that traversed the mudflats at the edge of the harbor on enormous timber pilings. A T-shaped section crossed the seaward end of the dock, allowing ships to pull alongside and parallel park. A warehouse at the T sheltered incoming and outgoing cargo from Valdez weather.[71]

Longshoremen, truck drivers, and a mechanic fought wind and rain to keep freight moving on and off the dock. A crew of drillers, engineers, lever men, firemen, crane men, and a dredge master maintained the harbor itself, keeping it free of mud and debris.[72] Men and women traveling from or to interior destinations, primarily Fairbanks, crossed the dock in both directions. They shared the space with freight coming in and headed for those same destinations. People coming to the Valdez dock, or people and freight moving into the interior from it, had one route to follow—the Richardson Highway.[73]

The Alaska Road Commission (ARC) kept a crew of twenty-seven in Valdez. Tractor and truck drivers, graders, mechanics, carpenters, demolition men, and men who operated compressors and jackhammers lived in and worked out of Valdez. They dealt with problems unlike those facing any other road crew in the world.[74] Just twenty miles out of Valdez, the Richardson Highway suddenly climbs into the Chugach Mountains at Keystone Canyon, rising three thousand feet in just a few miles to the infamous Thompson Pass. In a typical year, the snowpack in Thompson Pass accumulates to more than fifty feet—the height of a four-story building. In October every year, the ARC closed the pass, and travel through Valdez came to a halt. But Valdez itself accumulated a snowpack of five to seven feet, so after October the ARC crew turned its focus to keeping the streets of Valdez passable.[75]

Each May the ARC raced to clear a lane between gigantic snowbanks through Thompson Pass so that they could reopen it. At the same time, thousands of tons of snow began to melt. Meltwater cascaded down off the Valdez Glacier. The glacier thundered and groaned as ice melted and shifted. More meltwater cascaded out of Thompson Pass through Keystone Canyon heading to the ocean. The dirt and gravel of the Richardson Highway dissolved into muck. And every year at least a few of its rough timber bridges washed away.[76]

On May 31, 1941, the *Fairbanks Daily News-Miner* reported the first trip of the season, "made by a passenger car owned by the Alaska Motor Stages. George Edgecumbe, proprietor, his wife and bus drivers Nick Immel and Art Tucker . . . [made] a test run to make sure the Richardson was passable from Fairbanks to Valdez. The Highway passed the test. The first southbound run with passengers would happen on June 4."[77]

For seven months every year the Richardson saw a steady stream of traffic to and from Valdez. A special breed of truck drivers negotiated Keystone Canyon and Thompson Pass to deliver freight to the interior. Several companies like Alaska Motor Stages carried passengers in both directions in small buses and station wagons.[78] Fairbanks lies 362 miles north from Valdez. For passengers who didn't want to make the journey in one jump, roadhouses along the way offered rough accommodations and meals. At the Tanana River two legendary roadhouses—Bert and Mary Hansen's and Rika Wallen's—bracketed the river, one at each end of the ferry crossing. The little town of Big Delta emerged there.[79] Travelers and freight jammed Valdez each summer, and her citizens scrambled to serve them. The citizens worked virtually every daylight hour—and an Alaska summer offers a lot of daylight.

In October Valdez became a sleepy little town and Valdezians turned to surviving the winter months. In 1941 the *Fairbanks News-Miner* reported, "The last bus of the season left Fairbanks for Valdez on October 9, 1941. Passengers embarked on the SS *Aleutian* bound for Seattle." Mail came through Valdez in both directions. Airplanes supplanted dog teams for carrying mail as bush pilots and an airstrip came to Valdez. By 1941 bush pilots like Bob Reeves provided transport to and from the interior, even in winter. Their planes carried mail, small cargo—and made emergency rescues.[80] For instance, on November 10, 1941, the *Nanaimo Daily News* reported the following: "Gales of estimated hurricane velocity whistled through the Thompson Pass summit today grounding a rescue plane which yesterday brought Mark Nielsen and Mr. and Mrs. Ward Clay, an expectant mother, from the Territorial Road Commission relief cabin at Mile 35 where they had been marooned with six other persons for five days." But airplanes offered a tenuous connection at best. The difficulties and dangers of flying over Alaska's frigid mountains kept the number of successful bush pilots to a very few and turned men like Reeves into legendary folk heroes.

Valdez headquartered Alaska's Third Judicial District, including a courthouse and a jail. "Federal prisoners convicted of lesser crimes are sentenced to servitude in Valdez. There are few guards, and none are armed." The locals have a "stock jail story that if the prisoners aren't home by nine o'clock they are locked out for the night."[81] The city also had a dam and a small hydroelectric plant.[82] Blanche Nason, a missionary, ran a small orphanage in a donated home on Alaska Avenue, and in 1941 the city built a small hospital next door to the orphanage.[83] Alaska Avenue also sported two small hotels—the Valdez Hotel and Hotel Nizina—a bank, three grocery stores, a barber and beauty shop, and Ben's Beer Parlor.[84] Fishermen worked Prince William Sound, and the local cannery gave the town a distinctive odor. Valdez also had two newspapers.

Margaret K. Harrais, US commissioner in Valdez in 1941, vividly remembered the impact of Pearl Harbor and the war on the little town. Everybody stayed glued to the radio, anxious for news. If the war came to Valdez, the only escape would be the Richardson Highway, which would be hazardous in winter. She also remembered the army's arrival: "I did not know there was so much powerful machinery in the world."[85] A pure Alaska product, subarctic Valdez resembled no other small town anywhere on earth.

CHAPTER FIVE

Getting Off the Ship

Valdez Harbor offered no harbor pilots. The USS *David Branch* dropped anchor on April 29, 1942. On April 30 her captain, forced to an unassisted docking, managed to ram her bow into and through the end of the Valdez dock.[86] The grinding crash brought citizens of Valdez running. No matter, Public Roads Administration contractors would rebuild the dock later; the *Branch* tied up to its remains. Colonel Whipple's supplies and trucks would come. Right now, he had to get the road-building equipment and his twelve hundred soldiers off the ship, find them places to sleep and eat, and then get them on their way up the Richardson Highway to Slana. And he had to do all of that without having his Black soldiers come in contact with Alaskans.

On the morning of April 30, 1942, Whipple's regiment was poised to crowd the dock and overwhelm the little town. Anything or anybody coming off the *David Branch* came to the narrow wooden dock and the warehouse, traversed the long dock to where it turned into Alaska Avenue, and followed the avenue past the frame buildings of Valdez to intersect the Richardson Highway (see Photo 2). Snow such as the young Black Southerners had never imagined covered every inch of that path.[87] At the end of April, snow had pretty much stopped falling on Valdez. But the snow that had fallen all winter was piled everywhere. Through the winter of 1941–1942, the Alaska Road Commission crew plowed, shoveled, and hauled, removing snow from streets and sidewalks. Removed snow had to go somewhere, and it accumulated in enormous heaps on every unoccupied space in town. Wood smoke and coal soot stained the crusted piles brown and black. Streets and sidewalks resembled narrow canyons. Snow from sidewalks on one side and Alaska Avenue on the other was piled into a massive wall between them. From the sidewalk, one could hear street traffic but couldn't see it. The snowbanks compressed the already narrow avenue to barely one lane.

When a company of approximately 170 men, each carrying two barracks bags, came off the dock, slipping and sliding on the unfamiliar and slick surface, they filled Alaska Avenue. To make matters worse, the *David Branch* had carried a few small trucks. If men learning to walk on packed snow and ice had a problem, men learning to drive on it had a much bigger one. Vehicles came very, very slowly off the ship along the dock and made their way gingerly out on the avenue, fitting themselves into the mass of moving troops. And, of course, the local population needed to use their avenue.

From the deck of the ship, when a soldier got to the deck, he saw the crumpled dock and a warehouse with a big sign. "Valdez," the sign proclaimed: "Terminus of a Great Scenic Road, The Richardson Highway."[88] From the warehouse the dock stretched across dirty, oiled water to the town, a cluster of frame buildings. The Valdez Glacier towered behind the buildings.

Haywood Oubre had grown up in New Orleans and brought an utterly unique personal history to the 97th. He graduated from Dillard University as the school's first art major. He went on to graduate school in art at Atlanta University and did "creative touches" for a new student union at Tuskegee Institute. Oubre's draft board contacted him in April 1941: "They told me I'd be jailed if I didn't show up in New Orleans the next day."[89] On April 30 Oubre came to the deck of the *David Branch* and looked out across the tiny town at the massive snow-covered mountains that surrounded it, at the giant Valdez Glacier. Years later he remembered his reaction: "When you first behold the beauty and nature in Alaska, you are overwhelmed. It was in April. The snow was on the ground. I had my parka on, and I said, 'Praise God, I've never seen a landscape so beautiful.'"[90]

The view from the deck impressed every single man. But the frigid temperature and especially the snow impressed most of Oubre's fellows more than the scenic beauty. Staff Sgt. Clifton Monk grew up in Newton Grove, North Carolina, and learned to operate heavy equipment at Eglin Field. He reacted to the snow at Valdez, saying, "It looked like hell on earth."[91]

On April 30 the Headquarters and Service (H&S) Company, Whipple's staff, came off the ship first. Oubre and the others hoisted their barracks bags and filed, one behind the other, through the maze of hatches and companionways to the deck. Emerging, they got their first look at Alaska and a breath of fresh air after days in the putrid hold. In single file they moved in fits and starts. The men at the head of the line picked their way carefully down to the dock and over its slippery planks. Slowly they crowded between the cliffs of snow along Alaska Avenue and made their way to the airstrip just across the Richardson Highway.[92] While the H&S troops worked their way up and off the ship, the soldiers of Company E assembled behind them in the hold, waiting with their barracks bags. As the H&S soldiers finally cleared the deck above, Company E troops began to move up behind them. Stopping, starting, mostly waiting, they made their way, one at a time, down to the dock and took their turn between the snowbanks on Alaska Avenue.[93] The hold of the *David Branch* held 1,200 men and their gear. Morning turned to noon and then afternoon. The H&S troops had departed the ship, but the 160 soldiers of Company E were still working on it. The traffic jam kept getting worse. Before long, waiting soldiers and equipment outnumbered moving ones by a wide margin. It took Company E most of the rest of the day to get clear of the dock. Most of the regiment remained in the hold. Clearly 1,200 men and their gear wouldn't come quickly off the *Branch*. In the end, it took four full days.[94]

Even as they worked their way off the old troop ship and out between the snowbanks, the next problem loomed: Where could they go? Fenced into the little city by the surrounding mountains, they didn't have many choices. Colonel Whipple not only had to get his men off the ship, but he also had to get them out of town. For the moment, he and his H&S

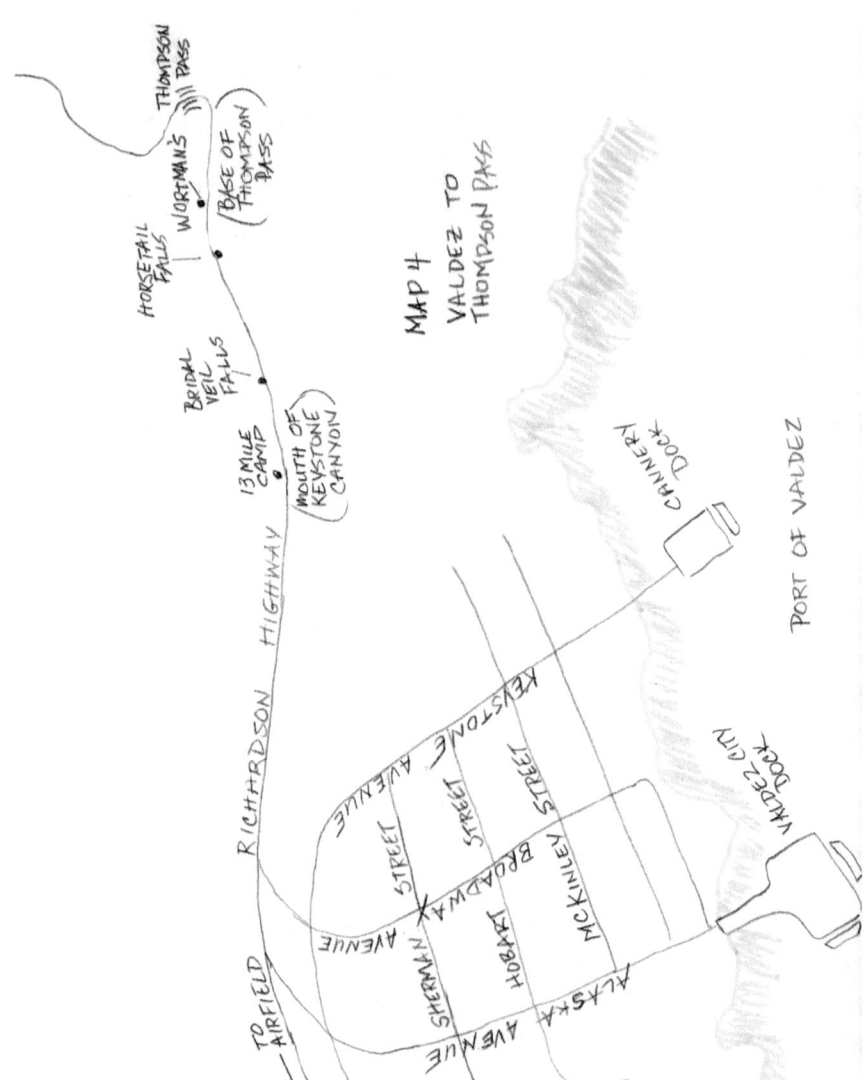

Map 4 Valdez to Thompson Pass Alaska

Company could sleep at the airfield. However, on the first day of unloading, April 30, Company E didn't stop at the airstrip. Emerging from the traffic jam, turning right onto the Richardson Highway, gradually finding their footing, and getting better at negotiating snow and ice, they walked out between even taller cliffs of snow on the highway. In two columns, one on either side of the narrow road, they trudged away from the Alaskans. Occasionally glancing up at the dramatic landscape, they mostly concentrated on the man in front of them and their slipping feet. Five hours later, thirteen miles out, Company E found a relatively flat area that could accommodate a camp.[95] They fell out and began scraping snow out of the way and pitching tents.

Four days later, on May 3, when the last company, Company B, cleared the ship, they walked directly out to join Company E at the thirteen-mile camp.[96] Thad Bryson, walking out that day with Company B, had come a long way from his experience with the Tuskegee Airmen at Eglin Field. He had come even farther from rural North Carolina. Over the next few days, the thirteen-mile camp blossomed into a tent city. On May 10 companies C and F swelled its population to nearly seven hundred soldiers.[97] Four companies filled the thirteen-mile flat spot to capacity. The rest of the regiment would have to stay at the airstrip for a few more days.[98]

Colonel Whipple needed to move his line companies out on the Richardson to Slana, his assigned starting point. But on May 10 two factors prevented that. First, his companies didn't have trucks. They couldn't march inland; it was way too many miles. To get to Slana they would have to convoy, and a convoy requires trucks. The colonel had Captain Parsons in Seattle working on that.[99] Second, just six miles out from the thirteen-mile camp, at Wortman's, the Richardson Highway climbs three thousand feet through Keystone Canyon to snowbound Thompson Pass. As they did every year, the Alaska Road Commission had closed the pass the previous fall. Now, in May, they worked frantically to clear the snow and open the pass. The men of Company E moved up to camp at Wortman's and pitched in to help.[100]

But the 97th would go nowhere until the pass opened. Wherever they found to settle during that first week in Alaska, soldiers had to pitch and then sleep in canvas tents. They had to cook and eat. When a company arrived at a bivouac, the company commander showed his four platoon leaders where their tents would go, and with his first sergeant he would find a central location for his headquarters, his officers, his kitchen, and the mess tent. The platoon leaders and sergeants would disperse, taking their men to their assigned area, directing each squad leader to the location of his squad's ten-man tent. The soldiers would then go to work—clearing snow away, laying out the canvas, and driving pegs through loops into dirt to hold it down. But in early April, Alaska dirt is the consistency of a brick. Some soldiers figured out creative, if makeshift, solutions. A man could tie the canvas to something in place of a peg. The tent would sit crooked, but it would sit. Some gouged holes out of the brick earth, inserted the peg, poured water into the hole, and waited for it to freeze. When the time came to eat, the company mess offered little except gray boxes of rations—suspicious concoctions in green cans, cold and coagulated. Fires sprouted, and men figured out how to heat the cans. Finally it was time for the exhausted men to go into the tents and climb into their sleeping bags or bedrolls. Sergeant Monk remembered, "I had one blanket. My buddy had one blanket and an army jacket."[101] In early April Valdez enjoys an average high temperature of thirty-six degrees, but at night the average falls to twenty-four. The men lay shivering in the dark, trying to sleep, and wondering what the hell they had ever done to deserve Alaska.

Thankfully, the cold didn't last long. Over the three weeks that Thompson Pass kept the regiment in place, winter eased its grip. But Alaska had more treats in store for the young Black Southerners. Warmer temperatures melted the snow and thawed the icy ground. Vehicles on the Richardson Highway no longer slipped and skidded on ice and snow; now they slogged through and sank into sucking mud. As the ground everywhere softened into muck and temperatures warmed, mosquitoes swarmed. Sergeant Monk put it simply: "The mosquitoes like to have eat me up."[102]

Staff Sgt. William Griggs had grown up in Baltimore and eventually came to the army and the 97th. In April he had scored a leave and went home to get married. Between his marriage and honeymoon, a telegram urgently summoned him back to Eglin Field. Sergeant Griggs remembered the mosquitoes well: "There were one thousand per square yard. We slept under netting and wore WWI hats which were covered with netting to protect our faces."[103] Haywood Oubre remembered them too: "You had mosquitoes that dive-bombed you. They'd dive like the Japanese with the dive-bombing. They'd dive and hit you."[104]

Stuck in their tent cities, waiting for the pass to open and their trucks to arrive, wishing for letters from home that weren't getting to them, the soldiers had little to do. Morale suffered. Offering a time-honored army remedy for boredom, officers and NCOs conducted drills. Soldiers turned out to march and maneuver.[105] And they practiced. At the thirteen-mile camp, Lieutenant Mason's troops built a log "demonstration" bridge.[106]

Captain Parsons had trucks coming out of Seattle and up to Valdez. When Thompson Pass finally opened, all hell would break loose—no more marching and maneuvering and building demonstration bridges. The men would move on to the reality of their incredibly difficult project.

CHAPTER SIX

Roadblocks

While the soldiers in Valdez struggled to get off the *David Branch*, Captain Parsons, back in Seattle, struggled with the army's confused system of logistics and supply. On May 2 he received, of all things, eight boxes of band instruments—pretty certainly the last thing the regiment needed.[107] He wrote to Abbie that day: "Things have been coming in plenty fast now, but not going out so fast."[108] While he was writing the letter, a telegram from Whipple interrupted: "Before I could finish that sentence, I received a wire from 'Old Grandma' wanting trucks, trucks, trucks. He'll get a few but damn few because trucks are deck loads and the boats are all small."[109] "Old Grandma?" In April, as Whipple had taken charge and dealt with all the issues in getting from Florida to Alaska, his personality and leadership style had emerged quickly. Those things mattered a great deal to his subordinates. And to at least one subordinate officer, Whipple had emerged as "Old Grandma."[110]

On May 3 the soldiers in Valdez had cleared the ship, and Whipple fired off an urgent telegram to Parsons: "Transportation shortage critical." He noted that their cargo contained only three 1½-ton trucks, three 2½-ton trucks, and one 4-ton truck. "This transportation insufficient to support regiment on march . . . cannot leave vicinity Valdez."[111] Getting themselves off the ship, the soldiers could march to the airstrip and to the thirteen-mile camp. When Thompson Pass opened so they could get out on the Richardson Highway, they would travel 190 miles to Slana (see Map 5 on page 58). They couldn't march that far. They would travel by convoy in trucks. So "Old Grandma" demanded trucks. And as trucks arrived in Seattle, Parsons scrambled to find ships, barges, and any other kind of vessel to carry them up to Valdez. When he found a vessel, he crammed as much onto it as humanly possible. On May 5 he wrote to Abbie, noting the amusement of the Seattle longshoremen at his order to load jeeps into the back of any truck big enough to hold one—usually the army's ubiquitous $2^{1}/_{2}$ ton "deuce and a half"—before loading the trucks onto the ship.[112] The army poured equipment into Seattle. The other regiments on the Alaska Highway project, as desperate for equipment as the 97th, reached to grab what they needed. In a May 5 telegram, Parsons warned Whipple that the transportation department had diverted a shipment from Valdez to Skagway. "Only a telegram from you to Commanding General Alaskan Road can save same."[113]

Up in Valdez, food, weather, mud, mosquitoes—seemingly everything—conspired to make life miserable. To make matters infinitely worse, the soldiers, to a man, missed home and family. They had been away from their loved ones for a very long time, and when they left Florida, mail service had effectively stopped. On May 9, "Old Grandma" wrote to Parsons: "Please ascertain why mail accumulated to April twenty-ninth not shipped on SS Alaska and arrange for prevention of recurrence."[114]

Captain Parsons's primary efforts bore fruit, and by the middle of May, ships and barges carrying trucks began pulling in and tying up to the crippled Valdez dock—first a trickle and then a flood. The soldiers of the regiment waited, scattering from the Valdez airfield to the thirteen-mile camp to foundations of the old Wortman's Roadhouse. Trucks arriving at the dock had to get off the ships and out to the men. And they had to do it quickly, because Parsons had more ships carrying more trucks headed for the dock.

The army, wanting Whipple's Black soldiers out of Valdez, had assigned a company of Alaska National Guardsmen to Valdez to unload ships and move material inland to the Black soldiers. But the National Guardsmen hadn't arrived yet.[115] When Whipple's precious trucks started coming in, he assigned thirty Black soldiers from Company B to unload them and get them out to the companies. The men from Company B scrambled aboard each arriving ship, unchained the deck-loaded vehicles, and fired up the roaring, whistling diesel engines. Trucks and jeeps rumbled down to the dock, along its length, across the mudflats to Alaska Avenue, and then moved out in file on the avenue to the crowded airstrip. At the airstrip men from the line companies met the Company B men and took over to drive them out on the Richardson Highway.[116] Like the docks, the Richardson had never seen such traffic. And it came at the worst possible time of year. A rough, one-lane gravel road at its best, the highway required constant maintenance. In May the spring thaw turned mountains of snow into water, and stretches of the highway dissolved into rutted muck that heavy vehicles churned into a quagmire. Soldiers fell out to fix and maintain them. And the trucks kept moving.

Pulling out of the airstrip and turning onto the Richardson, a driver faced a flat stretch of dirt road bordered by a few frame buildings. A pole frame supported a wooden sign reading "Richardson Highway."[117] He rumbled along for a few miles to his company—at the thirteen-mile camp or at the old abandoned roadhouse called Wortman's. His eye might follow the line of the road farther, to an ominously looming range of mountains, but for the moment snow still plugged the pass through them. The companies, gaining mobility, waited. The pass opened on May 20, and the soldiers of Company D convoyed out that day, piling into the back of trucks for the ride. They rocked and bounced on their wooden benches as the trucks moved slowly up Keystone Canyon. Through the arched opening in the back of the canvas truck covers, they saw the treacherous canyon and the remaining mountains of snow at the pass. Beyond the pass the trucks rumbled through the boredom of a long day. Getting accustomed to the rough ride, some of the soldiers occasionally dozed off, but only briefly. They talked among themselves—about home, about Alaska, and about what lay ahead of them. The trucks negotiated the long, difficult road, forded streams, and slogged through slick mud. The Alaska sun still shined high in the sky when the trucks stopped in the

58   A DIFFERENT RACE

evening. It would take the soldiers a lot more time to get used to Alaska's endless days without nights. They wearily climbed down from the trucks and milled around a bit, stretching to shake out the stiffness of the ride. Bellowing sergeants prodded them to unloading equipment, to pitching tents and other tasks.

On May 20 the soldiers of Company D stopped at Tonsina, fifty miles out.[119] Six days later the soldiers of Company C followed them through the

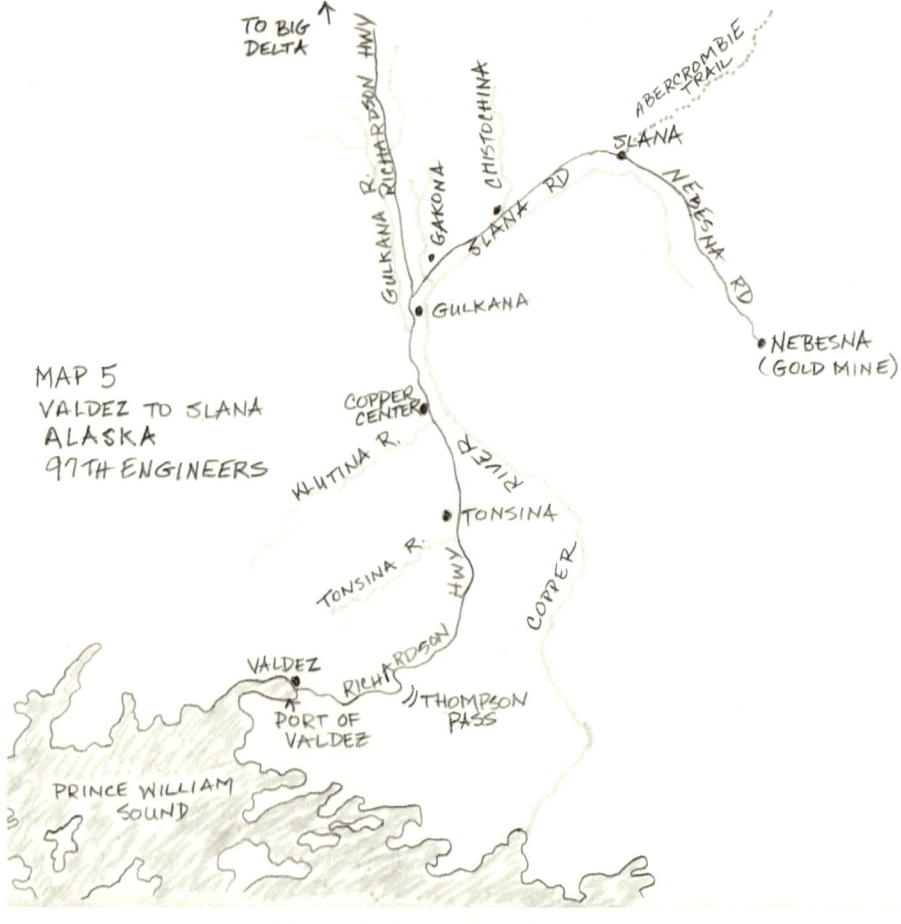

Map 5 Valdez to Slana Alaska

pass and moved past Tonsina into the lead.[120] Their trucks sloughed and rumbled 190 miles, averaging barely 13 miles an hour, all the way to Slana, where the state-maintained road ended altogether. At long last, Colonel Whipple had soldiers at his starting point.

Pvt. Major Banks didn't ride the trucks to Slana with his buddies in Company C on May 26. Six days earlier, on May 20 he had reported for sick call and the medics sent him in the opposite direction—back to the little hospital in Valdez.[121] Banks grew up in New Canton, Virginia. He didn't enter the army until January 1942, so he came late to the 97th, part of its hurried March expansion from battalion to regiment at Eglin Field. The army routinely vaccinated soldiers against yellow fever and had vaccinated the men of the 97th Battalion long since. They vaccinated Banks and the newer soldiers at Eglin Field in late February and early March.[122] Although the army didn't know it, during March 1942, in the frantic pace of vaccinating soldiers and shipping them overseas, they had vaccinated the men with contaminated serum. Two months later, in May, thousands of soldiers all around the world came down with serum hepatitis.[123] That month the army sent Private Banks back to the hospital in Valdez, where he was diagnosed with jaundice and progressive, acute atrophy of the liver. In truth, however, Banks had serum hepatitis.

As the soldiers moved, company by company, farther away from Valdez, Parsons's ships and barges continued to arrive at the dock. But now they carried bulldozers and other heavy equipment. For the young, relatively inexperienced drivers and operators, the trip out to the waiting companies at Slana got a lot longer just as it got a lot more interesting. A heavy bulldozer with caterpillar treads, sometimes towing a grader, rumbled over the still muddy Richardson Highway at a stately four miles an hour, past the quickly emptying thirteen-mile camp, the foundations at Wortmans, and into Keystone Canyon.[124] Entering Keystone, the dozer's young Black caterpillar operator, known as a "catskinner," found himself in a different world. Towering rock cliffs, punctuated at intervals by cascading waterfalls—Bridal Veil Falls and Horsetail Falls—closed in on him from both sides. He and his dozer climbed the narrow dirt passage cut into the cliff on his left, and as he did so, the cliff fell away on his right, a precipitous drop that went from hundreds to thousands of feet. Periodic ruts and washouts narrowed the road to barely more than the width of his dozer, and his right-side tracks ground over crumbling dirt right at the edge of the cliff.[125]

At the top of his climb, the tractor rumbled through Thompson Pass. The civilians of the Alaska Road Commission and the soldiers of Company E had cleared a path through the pass, but snow still towered four stories high on both sides of the road. Occasionally a section of the snowbank would collapse into the path, blocking it for a few hours while soldiers and civilians scrambled to clear it.[126] Beyond the pass the road descended past the Worthington Glacier. The glacier's surging meltwater had washed out some of the old timber bridges, and the Alaska Road Commission rushed to replace them. But even intact bridges couldn't support twenty-three-ton bulldozers. With the heaviest loads, the soldiers had to bypass the bridges and ford the rushing water.[127] Glacial melt swelled the streams, filling them to, and sometimes over, their banks. The rushing water—sometimes shallow, sometimes deep, changing depth abruptly, always frigid—rolled through a streambed with enormous power. A catskinner would carefully work his roaring dozer down the bank and out into the water, aiming upstream at an angle, knowing the current would push him downstream. Sometimes he made it to the other bank; other times the dozer sank into the muddy bottom or simply "drowned" and went silent when the water reached the engine. Then the catskinner sat, trapped, on his steel seat in midstream while his buddies figured out how to tow him out. The troops got creative with tow cables, trees, and other dozers.[128] Although they were now way behind schedule, the soldiers of the 97th and their equipment kept moving toward Slana.

CHAPTER SEVEN
Lytle and Green

Generals Sturdevant and Hoge dispatched the segregated 97th to Alaska because they could get them there quickly, but they expected little from the Black soldiers. In his March memorandum to the PRA, Sturdevant barely mentions the 97th, making clear his assumption that civilian contractors would build the road in Alaska.[129] The soldiers would come quickly, plunge into the woods, and start a rough pioneer road; civilian contractors would flood in behind them to accomplish the bulk of the mission. The army can dispatch soldiers, organized into military units with equipment more or less in hand, relatively quickly in an emergency. Soldiers in wartime face danger and endure hardship. Civilian contractors take more time to get organized, recruit workers, and get to the job. Civilians do not expect to face danger or endure unreasonable hardship. The generals' plan to get the 97th through Valdez and up to Slana had, to say the least, not worked as they expected in May. Getting civilian contractors in to surround and supplement the efforts of the Black soldiers didn't work as planned either.

The United States had begun working on Alaska's defenses even before Pearl Harbor, using civilian contractors. In early 1941 the Civil Aeronautics Authority had awarded contracts to two Iowa contractors, C. F. Lytle Company and Green Construction Company, to build airfields in Alaska, including one at Big Delta. When the PRA went looking for a management contractor to put together and supervise the network of civilians to work with the 97th, they found Lytle and Green already in place.[130] Negotiating contracts with the two companies took time, and passing time wreaked havoc on the generals' plan. In early May 1942, even as the soldiers of the 97th struggled alone, without contractor support, to get past Thompson Pass, away from Valdez and out to Slana, the PRA finally signed a joint contract with Lytle and Green.

Over the next two weeks, the "Iowa Expeditionary Force" loaded bulldozers, scoop shovels, graders, and trucks; chained them down on railroad cars; and shipped them north, some to the Port of Seattle, some to the Port of Prince Rupert, British Columbia.[131] But at Seattle and Prince Rupert, the equipment piled up behind a serious bottleneck. Equipment bound for Valdez had to negotiate the Gulf of Alaska, and contractors struggled to find vessels to carry it. On June 7 fifty carloads of equipment waited in the yards at Prince Rupert.[132] Each contractor employed a core of skilled and experienced men—catskinners, carpenters, mechanics, crane operators, foremen, and superintendents.[133] But to do the job in Alaska they needed more men than just these, and they scrambled to recruit laborers "from Cumberland to Independence, from Cedar Rapids to Hawarden and a dozen other Iowa communities."[134] Generous contracts with the PRA allowed civilian contractors to offer very high wages, and they had little problem recruiting laborers. In a letter to Abbie from Valdez, Captain Parsons noted, "Truckers are getting $25 to $30 a ton for a 200-mile haul and drivers are getting from $400 to $700 a month." And it wasn't just truck drivers. "Café workers make more than Army officers," Parsons continued.[135] Privates in the 97th, however, worked for just $21 a month.

In June, while equipment waited at Seattle and Prince Rupert, a flood of civilian workers made their way north. Twelve hundred men served in the 97th, and in the end an additional twelve hundred men worked for Lytle's and Green's contractors in Alaska.[136] A few of the civilians made their way to Valdez the way the soldiers did—north from Seattle by ship—but most traveled by rail to Edmonton, Alberta, and then flew north to Big Delta, Fairbanks, and Nabesna.[137] The workers assembled in Iowa towns, said tearful good-byes, and boarded trains. Iowa newspapers celebrated their heroic passage. Under the headline "Build Alaska Highway" and the subhead "Burma Road of North America Is Being Rushed," the *Iowa Globe Gazette*, out of Mason City, reported: "One hundred thirty-five working men of E. M. Duesenberg, Inc. and T. L. Sears Construction Company . . . have embarked for Alaska. Men from all points of Iowa and several other states also have left. They will be engaged in the construction of 150 miles of the transcontinental Alaska-Canadian highway."[138] As with the *Globe Gazette*, none of the flood of newspaper stories found it necessary to mention the 97th or the Black men who already labored in Alaska. The generals had hidden them well, at least from the lower forty-eight.

The young Black soldiers had not written letters home from the crowded rail cars or the stinking hold of the *David Branch*. Civilian workers, however, did write letters as they traveled. Don Garlock wrote of a special train to St. Paul, Minnesota, and supper in a reserved dining room at Union Depot. North from St. Paul into Saskatchewan, the railroad provided sleeper cars. Not everything went perfectly for the civilian travelers, though. Garlock recorded: "The result was catastrophic as there were only 113 berths for 191 men, so we slept two to a berth. Of course, there was little sleeping done. In fact, I dozed with the tingling of silver in my ears—poker game in the next berth."[139]

Army barracks housed the traveling civilians for a time at Edmonton. Max Smith "had a bath last night and the first good sleep since I left home." Max had breakfast at the Salvation Army canteen: "Ham and eggs with toast, coffee, and pie for 40 cents."[140] Waiting to fly north, the men had time to kill. "Ray Van Buskirk enjoyed a few hours on the golf course. . . . Don Garlock and others took in Deanna Durbin's movie 'It's a Date' at the Legion Club."[141] "There was time enough, too, to organize a pretty good baseball team." And some of the men organized a fishing expedition. Their adventure got a lot of publicity back home.[142]

Finally, at the end of June, civilian workers began piling into planes for the trip north. Forest fires raged north of Edmonton, and the planes flew through heavy smoke. Max Smith wrote, "I am writing this letter from a plane 15,000 feet in the air going 165 to 185 miles per hour somewhere over the northwest Canadian Wilderness. . . . The smoke was so bad we couldn't see the ground."[143] Willie Comstock wrote, "[We] got on a big army transport and flew 1600 miles in 12 hours with two stops. . . . Thousands and thousands of acres of timber and underbrush going up in smoke. . . . We flew at 9,000 ft. and the smoke was so thick that we couldn't see anything else and it came thru our ventilators."[144] Nervous, airsick, but on an exciting adventure, the civilians made their way north.

The work the generals' plan assigned to Lytle and Green would center on Gulkana, at the point where the Richardson Highway branched northwest toward Fairbanks and the road to Slana branched northeast. Lytle and Green leased a "two-story, peeled log structure," a roadhouse called the Gulkana Lodge, and made it their headquarters. Some of the workers flew into Fairbanks and made their way down the Richardson to Gulkana. Others flew into Buffalo Center, a grassy field near Big Delta. They too made their way down the Richardson. Still others flew to an airstrip near an abandoned gold mine at Nabesna and made their way back on the old mine's access road to Gulkana. Wherever they landed, the civilians stepped off their planes into the vast emptiness of Alaska. And no one had planned for emptiness. Two hundred men who landed at Buffalo Center found no food, no place to sleep, just a big grassy field. Don Mathiason remembered, "After hours of waiting and still no help we realized we would have to fend for ourselves. We built campfires and huddled around them while eating corned beef sandwiches bummed from a nearby workcamp."[145] Max Smith's plane dropped him at Nabesna, or as he referred to it, "God Only Knows Where, Alaska." He hiked out to the road to catch a truck that would carry him to headquarters at Gulkana, but he got there too late and missed the truck. He hiked back to the airstrip and found that his luggage had gone on a truck "another way."[146]

As the workers made their fitful, halting way to Gulkana, a forest of tents sprouted around the lodge, including a large circus tent. Don Garlock wrote home, saying, "It was quite hot here Sunday, but Monday it rained all day and last night it froze ice and there was heavy frost all over. And me sleeping on a cot in a tent with no floor and no heat and 3 blankets over me."[147] The Black soldiers at the thirteen-mile camp back in April would have given a lot for the cot and the blankets. Most crucial for the generals' ill-fated plan was that the gathering at Gulkana didn't happen until early July, and their equipment still languished in Seattle and Prince Rupert. So through June and into July, the Black soldiers of the 97th continued to work into Alaska on their own. For better or for worse, whether the generals liked it or not, the troubled 97th remained in the lead.

CHAPTER EIGHT

Disorganization and Getting Started

Getting away from Valdez at the end of May and into June, army convoys followed the Richardson Highway out 120 miles to Gulkana. At Gulkana the Richardson angled off northwest to Fairbanks. The convoys left the Richardson there and headed to the northeast on a branch wagon road. Sixty-seven miles out on that road, the convoys reached the point where the Slana River flows down out of the Alaska Range and into the Copper River. There, at an abandoned roadhouse, at a place called Slana, all semblance of a road ended.[148] Back at the turn of the century, Cpt. William R. Abercrombie had built his trail from Slana through the Alaska Range and on to the Tanana River; it was a rough, dangerous trail meant for pack horses.[149] Unused for decades, it had all but disappeared. Unfortunately, the old trail still existed on General Sturdevant's and General Hoge's maps and in their imaginations. They blithely directed the soldiers of the 97th to start at Slana and replace 77 miles of Abercrombie's trail with a "rough pioneer road" through the mountains and on to the Tanana River.[150] Competent civilian contractors would come right behind and build an actual road. (See Map 3 in Chapter 3, on page 37.)

In June, of course, none of the generals' civilians had made it to Alaska. The soldiers of the 97th, starting their road building out of Slana, had no one to depend on but themselves. On May 26, while civilian contractors loaded equipment in Iowa, the soldiers of Company C arrived at Slana.[151] The soldiers of Company B joined them on June 8, and those of Company A arrived on June 12.[152] A tent city ballooned at Slana with half the regiment in place, ready to build a road. A location party under Lt. Joseph Raso headed out first to find and drive stakes along the old pack trail. Leading a pack mule with food, camping gear, and a few tools, Raso's party made its way north along the Slana River. On June 12 his party reached Carlson Creek, eight miles away from Slana. Eight days later Raso's men, working their way through the woods along the Slana River and approaching Mentasta Lake, came to a ravine. Locating a dead tree that had conveniently fallen across the ravine, they ventured out to cross on its trunk—a good idea for the men, but not so much for the mule. Halfway across, the mule slipped, straddling the log, and dropped on its belly, hee-hawing at the top of its lungs. Raso and his men had to remove the mule's pack and then drag the very unhappy animal across to where it could stand on solid ground.[153]

Behind Raso and his location party, the soldiers of companies A, B, and C swung into action. Lt. Walter Mason remembered, "We started the road about two p.m. Sunday. . . . I was standing on the side of the D8 Cat angle dozer when it made the first cut."[154] The monstrous Caterpillar D8 bulldozer, weighed twenty-three tons and could move over the ground at 5.8 miles per hour.[155] The army officially doubted the ability of Black soldiers to operate bulldozers. *The Corps of Engineers: History of Troops and Equipment* stated flatly that "Negroes lacked the sense of responsibility necessary for the care of equipment . . . and were slow to absorb instruction."[156] Out of Slana, though, proud Black catskinners worked the levers of the D8; its engines roared, spewing smoke, and its giant tracks ground through the sandy soil. Along the west bank of the Slana River lay the Slana sandhill, and that's where they started. The dozers tended to slip sideways instead of moving forward, but the catskinners quickly learned to deal with that, angling their cut into the hill.

The regimental adjutant, Cpt. Jack Doyle, wrote a report that described the work during those first days: "Starting from Slana, Alaska, with inexperienced operators, the cats started a side hill cut on the Slana sandhill. While the operators were still getting the feel of the dozers and graders and carryalls, the road followed the Slana River to Lake Mentasta."[157]

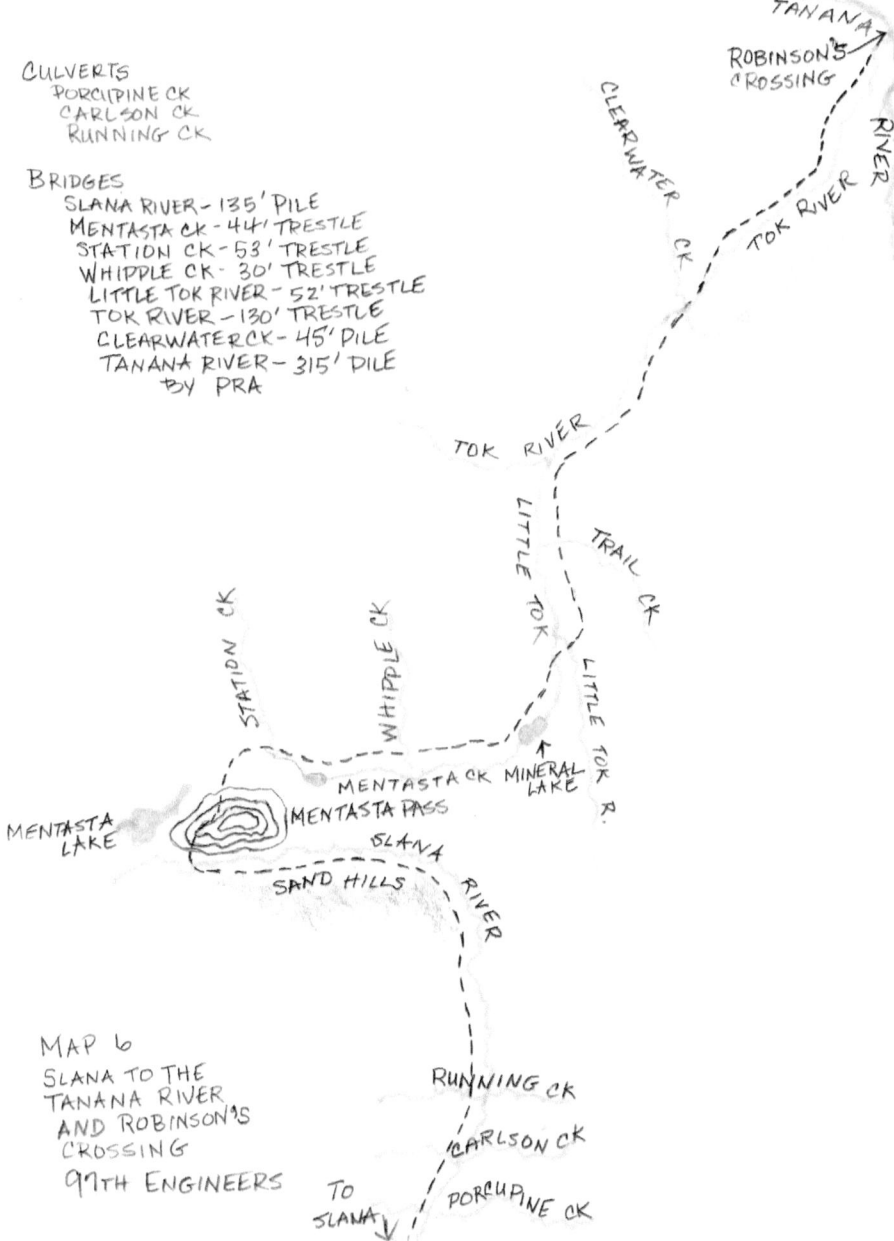

Map 6 Slana to the Tanana River and Robinson's Crossing

Doyle makes clear that despite the army's and the generals' dismal expectations, the inexperienced Black soldiers worked over the sandhills, gaining experience the hard way but very quickly. The work soon fell into a pattern: the soldiers of Company C led the way, blazing a rough one-lane road; those of companies A and B followed them, widening, upgrading, and grading.

By June 12 the soldiers had followed Raso's stakes with "a rough one-way road fourteen feet wide blazed for two and one-half miles."[158] On June 20 the soldiers of Company C reached Carlson Creek, eight miles from Slana. The regiment lucked into a deserted sawmill, put it into operation, and some of the young Black soldiers quickly acquired another set of skills. The men of companies A and B used the lumber to construct the regiment's first culvert and to bridge Porcupine Creek.[159] Half the regiment—in army parlance, the 1st Battalion—built road. But with most of General Sturdevant's and General Hoge's civilians still in Iowa, the 2nd Battalion, the other half of the regiment, labored in June to do work that the generals had confidently expected civilian contractors to do.[160] The dozers and trucks at Slana emptied fifty-five-gallon fuel drums at a breathtaking rate. The soldiers there needed food, spare parts, and myriad other supplies. The 2nd Battalion struggled to deliver material over the long road from Valdez. Worse, the crude bridges and rough surface of the Richardson Highway crumbled under heavy truck traffic. Other soldiers of the 2nd Battalion labored to keep the road passable.[161] A trickle of civilian contractors and equipment came to Alaska through June, gradually taking over material delivery and road maintenance, which allowed the soldiers of 2nd Battalion to gradually move out toward Slana. But the bulk of the civilians would not arrive until July.[162]

Maj. Saul Gordon commanded the 2nd Battalion. However, he, along with the company commanders and the other officers of 2nd Battalion, presented "Old Grandma" Whipple with a problem that the colonel was not equipped to solve.[163] So while the Black soldiers of Gordon's three companies labored as best they could to deliver material and maintain the road, they labored without supervision, direction, or coordination.

Gordon and his subordinate officers payed little attention to the soldiers or the work. Instead they bickered, politicked, and just plain avoided work whenever they could. Another commander might have grabbed control of the situation early on, fired or simply marginalized Gordon, clearly defined the officers' responsibilities, and established and enforced strict standards of performance. Fastidious, nitpicky "Old Grandma." simply had no idea how to delegate and enforce.[164]

A memo that Col. Lionel Robinson of 1st Battalion later wrote to all of the officers of the regiment makes clear the problems among Major Gordon's officers in June: "In some instances, I've seen it take longer to figure out some reason for not doing a thing than it would have taken to do it in the first place. . . . Many times I will find two-three-four-and five officers congregated where they were not needed at all. Day after day I see men in groups of fifty to seventy-five with no officers present."[165] He didn't stop there, adding, "I've seen an officer draped over the steering wheel of a pickup, sound asleep while he was supervising the work of several trucks, two tractors, and many men."[166]

Once Captain Parsons joined the regiment in Valdez, in his letters to his wife, Abbie, he made clear his opinion of Gordon, Gordon's subordinate officers, and "Old Grandma's" failure to address the problem. In mid-June Parsons escaped from Seattle, if not from his duties for "Old Grandma." His June 13 letter to Abbie described the last leg of his voyage, through a North Pacific gale, and reported his arrival in Valdez that afternoon: "Col. Whipple and H&S have headquarters here—but now that I'm here they are going to leave me as liaison officer." Whipple's 1st Battalion labored out of Slana, his second worked between Valdez and Slana. His headquarters temporarily remained in Valdez with 2nd Battalion. Now that his liaison officer Parsons had arrived in Valdez, Whipple could move north to Slana with the bulk of his troops. The liaison officer assignment had landed Parsons right in the middle of Major Gordon's squabbling subordinate officers.[167] Parsons explained to Abbie the missions of the regiment's battalions: "Gordon's gang is maintaining the road from here to the jump off place [Slana]. . . . [1st Battalion] has started there already with what equipment they have." The disorganization and mismanagement at Valdez headquarters horrified Parsons. Whatever might be happening up north from Slana, in Valdez chaos reigned. In the privacy of his letter, he wrote, "This is one hell of an organization—nobody knows who is who or seems to care . . . When they get a job, they just all start to do something with no organization, nobody knowing what the next guy is doing, going to do, etc."[168] On June 16 Gordon and two of his subordinate officers came to Valdez, and Parsons described for Abbie their attempt to get him involved in the bickering: Some of Gordon's subordinates worried about Parsons's future role in the regiment, worried that he might displace one of them. "[They] don't need worry because it would be mighty hard to get me in 2nd BN [Battalion] now."[169] Colonel Whipple and most of his H&S staff left to establish headquarters at Slana on June 17. The rest of the men of the H&S Company, fifteen enlisted men and five officers, along with the detachment from Company B, moved with Parsons nineteen miles to the encampment at Wortman's, at the entrance to Keystone Canyon.[170]

Enlisted Black soldiers struggled to live, work, and get the job done under the disorganized and self-serving leadership of the 2nd Battalion. Among them, Cpl. James Heard of Elberton, Georgia, was promoted to sergeant on June 16.[171] On that day Heard took charge of a squad of nine other young Black men and assumed responsibility for how they would live and work in Company F of Gordon's troubled battalion. Bad officers didn't shock Sergeant Heard or the other Black soldiers serving in Gordon's battalion. They had worked for white bosses back in the Carolinas and Georgia, quite often bad white bosses. Now they worked for bad white officers. The white men, like Gordon's officers, sometimes ignored you, left you to proceed with your work, and blamed you when things didn't work out. The young Black men knew to stick together, to respond as a group, and to respond very, very carefully. The white man, if he chose, could visit devastating consequences on the Black man. Black soldiers, sergeants and privates, mostly avoided white officers, kept their heads down and their mouths shut, and tried to get the job in front of them done as best they could.

Sgt. James Morton was one of those who moved to Slana with Whipple's headquarters on June 17. James had grown up in Locust Grove, Tennessee, and dropped out of school in the ninth grade. In January 1941, only sixteen years old, he had forged his grandmother's name on a permission slip so that he could join the army. He came to the 97th as a corporal at Camp Blanding. In October 1941, as a staff sergeant, he took over the regimental message center in the H&S Company.[172]

Sgt. William Griggs, who also moved to Slana with H&S on June 17, had learned photography from his father and had honed his skills at Lincoln University in Pennsylvania. He brought those skills to the 97th at Eglin Field, where he became the official regimental photographer. Griggs had special permission to make occasional trips back down to Valdez to buy film.[173]

Parsons's portion of the H&S company and the truck drivers from Company B struggled through June to get the remaining equipment and supplies from the dock, through Valdez, and out to the regiment: "I'm going up the road . . . after a bit to see how our big shovel which left about 2 hours ago is getting along. . . . Our drivers go day and night as it's light all the time."[174]

At the end of June, one small piece of General Sturdevant and General Hoge's plan to keep the Black soldiers away from Alaskans fell into place. A battalion of white National Guardsmen arrived in Valdez to unload equipment, thus freeing some of Parsons's detachment from Company B to rejoin their company.[175] Down to fifteen men and five officers at the Wortman's camp, Parsons lived with a rotating cast of Major Gordon's officers, sneaking down for a Valdez steak and a comfortable bed.[176] On June 30 Parsons wrote to Abbie, "One of the colored boys died early this morning and things are in quite a stir about this little camp." A month earlier the regiment had sent Private Major Banks back to the hospital in Valdez.[177] Banks had lingered for several weeks, getting progressively weaker, until the last day of June.[178] Parsons determined that Private Banks deserved the honor of a military funeral and set out to get him one. The citizens of Valdez objected; they didn't want a Black man buried in their cemetery. But Parsons would have none of that. In the end they designated a piece of ground just across the creek from the cemetery as a negro section.[179]

Parsons arranged for "a truck load of boys [to come] down to bury the chap. . . . We fixed him all up in a casket we got off of a boat. . . . Had a firing squad, bugler, military escort and everything plus about ten officers and twenty-five or thirty white soldiers from a nearby camp."[180] Ironically, those white soldiers came from the Alaska National Guard company that had come to Valdez to unload and transport supplies out to the 97th in order to keep the Black soldiers away from town.

CHAPTER NINE
Attack in the Aleutians

The advance of the Japanese war machine across the Pacific inspired the Alaska Highway project. But the soldiers and civilians who went north to build the highway, leaving the rolling catastrophe behind, struggled to keep up with news of the war. If few understood the complex geography of the Pacific, in early 1942 everyone understood that the Japanese, marauding through that geography, attacked wherever and whenever they pleased.[181]

Recall that the Aleutian Islands extend from mainland Alaska eleven hundred miles into the North Pacific. The island called Unalaska, site of the American navy base at Dutch Harbor, lies closest to the mainland. Adak, Kiska, and Attu lie at the other end of the chain, perilously close to Japan. Peaks of a submerged volcanic mountain range, the Aleutian Islands traverse the coldest, most turbulent portion of the North Pacific. Thick fog shrouds the treeless islands almost constantly, and williwaws—gale-force winds up to eighty miles an hour—come and go frequently and unpredictably. Isolated, inhospitable in the extreme, the Aleutians don't count for much. But they are, indisputably, American soil. And at Unalaska they are, indisputably, immediately adjacent to the North American continent.

In May 1942 Japan dispatched two fleets to the Aleutians: a carrier fleet to bomb and destroy the Dutch Harbor complex, and a troop transport fleet to invade and occupy Kiska and Attu. The Japanese, in other words, did exactly what the Canadian and American governments had feared they might do. On the night of June 2, 1942, the Japanese carrier fleet steamed through cold rain and icy fog less than 170 miles from Dutch Harbor. At 2:43 a.m. on the morning of June 3, they steamed out of the storm into the clear and launched thirty-five warplanes—bombers and fighters. Half of the pilots lost their bearings, couldn't find Dutch Harbor, and had to turn back, but fifteen planes made it through. At about 5:50 a.m. the eye of the storm passed over Dutch Harbor, clearing the rain and fog just as the Japanese planes descended into the attack (see Photo 4). They hammered the base and the harbor for the next twenty minutes American defenders launched puffs of flak into the sky. Machine gun tracers arced up from the ground. Two lumbering seaplanes managed to get into the air. But the Japanese bombed and strafed with relative impunity. They finished dropping their payloads into the churning smoke and flame boiling up from the ground, formed up, and flew away, convinced they had heavily damaged their target. Luckily for the Americans, they had not. Ignorant of the layout of the base, the Japanese pilots had engaged targets at random. They killed fifty Americans and destroyed a tank farm. But as the smoke cleared, vital facilities emerged unscathed. (See Map 2 in Chapter 3, on page 32.)

The weather had proved far more effective than the American defenders against the Japanese. Now, trying to find and fight the Japanese, Americans fought the weather too, spending June 4 in a frustrating, uncoordinated effort to find and sink the Japanese carriers. That afternoon Admiral Kakuji Kakuta launched seventeen bombers and fifteen fighters for a second assault on Dutch Harbor. This time they pounded an old ship, the *Northwestern*, which the Americans had beached to use as a civilian barracks. A steel building collapsed. The bombers destroyed one wing of the base hospital. Four fuel storage tanks went up in a gigantic explosion. On June 7 Admiral Boshiro Hosogaya, commander of the invasion fleet, landed troops on the island of Kiska. He landed more troops on Attu on June 9. An estimated twenty-five hundred enemy troops now occupied the western Aleutians. They would occupy American soil until a bloody invasion dislodged them in May 1943.[182]

In Alaska—indeed along the whole length of the highway—soldiers heard the news and wondered if the Japanese would come their way next. Mail service to and from the young soldiers, already spotty and intermittent, came to a complete halt for several weeks in June. Captain Parsons's letters to Abbie through June repeatedly report that he hasn't received any mail. Finally, on July 3, he wrote, "Your letter marked No 2 dated June 12 arrived yesterday. Was I glad to get it? You guess. It was my first letter in almost a month."[183] On his way to Alaska from Seattle in June, Parsons reassured Abbie in a burst of bravado that she shouldn't worry about Japanese attacks "in Alaska because it won't affect my gang. After we leave port of embarkation the japs won't be able to find us and if they do, they will wish to hell they had not because we have enough arms to outfit a Division."[184] But he also reported, "The morning after we got in . . . a sub was sighted off the mouth of this bay [Port Valdez] in which we landed."[185]

Ralph Green, leading the effort in Alaska for Lytle and Green, reported that the attack on Dutch Harbor had abruptly slowed shipping to Alaska.[186] Milton Duesenberg was the head of Duesenberg Construction, the company slated to follow the 97th north from Slana, which left Iowa on June 4. Traveling north through Prince Rupert, he encountered blackouts, air alarms, and patrol boats all along the coast. In his diary he recorded on June 9, "Prince Rupert had an air alarm today. I was ordered off the streets by MP's [military police]. The planes were identified as friendly."[187] Milton took a ship north on June 10. On June 12, between Ketchikan and Skagway, they passed a "US patrol boat and they trained the cannon on us but didn't shoot thank heavens."

Milton's brother, Warren, in Fairbanks on July 8, noted in his own diary, "War conditions much worse than people think. Army has complete control of all shipping. Eight bombers landed in here today. Planes are bringing back wounded from Dutch Harbor."[188] Sears Construction Company, assigned to build a second dock at Valdez, worked under the watchful eye of army "lookouts in jeeps on the mountains behind Valdez. The jeeps were equipped with red lights which could be flashed on and off to warn the civilians to douse their lights and turn off the machinery."[189]

In July, with their laborers finally arriving at Gulkana, the contractors struggled to get equipment there so that the men could go to work. The reaction to events in the Aleutians made their efforts even tougher. The bottleneck they had to push through lay at the ports. The navy patrolled the Gulf of Alaska and restricted shipping through it. And they couldn't find boats—that is, until they found the *Scottish Lady*.[190] Once a proud clipper ship, adorned with an elaborate female figurehead, the *Scottish Lady* had long since fallen on difficult times. The contractors in Seattle found her in July and the equipment bottleneck finally began to clear.[191]

The reactions and overreactions to the Dutch Harbor attack and the occupation of Kiska and Attu caused problems for both soldiers and civilians on the Alaska Highway project. But more than anything it ratcheted up the pressure on the road builders. For the moment, the Japanese had confirmed the fears that had propelled the builders north in the first place.

CHAPTER TEN
Contractors at Valdez

In July Lytle and Green's civilians flooded Gulkana and filled the big circus tent to bursting. The contractors had never operated in total isolation, and Alaska threw them a curveball. Owners and managers surely knew that Alaska didn't offer taxis, buses, restaurants, hotels, and medical clinics, but they hadn't thought enough about the implications of that situation. Consequences began to reveal themselves as soon as workers stepped off their planes at Fairbanks, Big Delta, and Nabesna and, finding nothing there, struggled in confusion to get themselves to Gulkana.

Concerns continued to surface when they arrived at Gulkana. A thousand men slept in shifts in the circus tent.[192] With no equipment in hand they couldn't work. But they could and did eat, going through groceries at a terrifying rate. Replacement groceries came slowly if at all.[193] Every so often, of course, workers got sick or were injured. Gulkana had no doctors. Milton Duesenberg wrote in his diary on July 9, "Took three men to the Army doctor after dinner. . . . We were informed . . . that the matter of doctors was brought up . . . but no decision had been reached."[194] In comparison, the 97th, being a military unit and inherently more self-sufficient, came somewhat better prepared for such isolation. The soldiers of the 97th slept in tents. But they had enough tents, so they didn't have to sleep in shifts. The soldiers didn't enjoy their food, but at least they had food. And the 97th came to Alaska with attached medical units.[195]

The civilian workers got filthy. The nearby Gulkana River offered the only facility for bathing or doing laundry. Max Smith wrote home, "I washed yesterday and today both. I guess I told you about taking a bath in the river, boy it is cold."[196] The contractors hustled to build showers, and Max wrote home again on July 26: "I really feel good today. I got up this morning about 9:30 went over and took a shower. With hot water, that is something, it is the first one since I left Edmonton."[197] The soldiers of the 97th got filthy too, but in the field the army didn't worry much about cleanliness. The soldiers wouldn't see hot showers anytime soon.

The contractors finally broke through the bottlenecks at Prince Rupert and Seattle in July. They assembled a motley collection of ships and barges, and their equipment began to move up to Alaska, first in a trickle and then in a flood. If Valdezians thought the departure of the 97th would end the convulsion of traffic at their dock, they soon learned different. Sears Construction built a second dock, and traffic in Valdez stayed convulsed—the new normal. Captain Parsons wrote to Abbie about the changes at Valdez: "This place has changed a whole lot lately. . . . The storage of supplies is quite a problem and even unused churches are being used as storage houses."[198]

The stripped-down *Scottish Lady*, now a barge to be towed by a tugboat, transported most of the civilian Iowa Expeditionary Force's equipment that came through Seattle. Lytle and Green had assigned Gus Ostermann's company to lead the effort to build the Alaska Highway from Big Delta south to a meeting with the 97th at Tanacross. Ostermann got to the *Lady* first. He personally supervised the loading of his equipment and then climbed aboard to travel along with it to Valdez.[199] From the Valdez dock, Ostermann's equipment had three hundred miles of Richardson Highway to traverse to get to Big Delta. Bulldozers moved up the gravel road at five miles an hour, and the Richardson crossed stream after stream. No one, as the men of the 97th had discovered in May, had intended the bridges on the route to support bulldozers. Forced to ford the streams, Ostermann and his catskinners resorted to the same desperate expedients the soldiers had been forced to use. They had to disconnect engine fans so that the current wouldn't push them into the radiator, and at least once Hans Ostermann "took a very cold dive in 35 feet of water in order to hook a cable to the drawbar of a stranded D8 Cat."[200]

E. M. Duesenberg Company's equipment followed Ostermann's up to Valdez. Of all the civilian contractors, Duesenberg would work closest to the 97th. Lytle and Green assigned them two of the three primary missions: upgrading the road from Gulkana to Slana, and then following the 97th out from there, maintaining and upgrading their pioneer truck road. At the beginning of July, brothers Warren and Milton Duesenberg had men in Gulkana and more coming. On July 18 a tug towed the *Scottish Lady*, loaded with Duesenberg equipment, to the Valdez dock.[201] Warren and Milton kept diaries, and Warren recorded that stevedores started unloading their equipment on July 20 and continued the laborious process through July 21 and July 22.[202] On July 23 the first Duesenberg drivers and catskinners left Valdez to take their turn at Keystone Canyon, Thompson Pass, and the long road north. Warren dispatched two D7 dozers that day, and Milton met them at the entrance to Keystone Canyon that evening. Two days later another vessel brought scoop shovels to the dock, and Warren headed down to supervise unloading and dispatching them.[203] Slowly, laboriously, through the end of July the Duesenbergs moved equipment north to Gulkana. With the logjams in Seattle and Prince Rupert finally broken, equipment for all the contractors flooded into Valdez. The Duesenbergs had lots of company on the Richardson.

The relatively experienced civilian operators and drivers documented their fierce struggle with the road. They wrote reports, histories, and even poems.[204] One can only imagine how the young Black men from the Carolinas and Georgia had experienced the trip a month earlier. They didn't write poems. But we do know that Col. Stephen Whipple, "Old Grandma," and his men had struggled. According to historian Heath Twichell, "Many of the 97th's problems could be traced to its commander. In the eyes of many of his junior officers, Colonel Whipple was a fussy, plodding nitpicker who lacked rapport with his troops."[205]

Through June a trickle of civilian contractors had arrived to start hauling supplies out of Valdez and maintaining the fragile roads. That had gradually freed up the soldiers of the 2nd Battalion. Company E had finally got away from Keystone Canyon and Thompson Pass on June 19, and Company D had finally left Tonsina. On June 23 Company F had escaped from Gakona. Moving into July, "Old Grandma" finally had his entire regiment in place at Slana.[206] But through July the leaderless soldiers of 2nd Battalion didn't accomplish much.

Major Gordon and his officers spent as much time as they could enjoying the food and accommodations in Valdez, leaving their soldiers to fend for themselves. Whipple left Captain Parsons and a few men from the headquarters company in Valdez through July, so Parsons observed the antics of Gordon and his officers from a front-row seat. "Old Grandma's" failure to bring Gordon and the others to heel infuriated him.[207] On July 5 he wrote to Abbie, "Gordon and Mrozek came in yesterday. . . . They found some poor excuse to get away from camp and came down here. It's true they can take a bath and buy a steak down here, but I'd give them my part any time. I think I could do without steak . . . and could rig up something for a bath without driving 200 miles for it."[208]

On July 12, itching to get out on the road, Parsons traveled north to visit Slana and see what was going on: "They are doing a good job and getting along fine. A lot of throat cutting etc. . . . If [Gordon] would fall off the side of a high mountain one of these days and grandma would just wander off, this thing could be worked out in much better shape."[209] On July 16 he shared a disturbing rumor with Abbie: "Smith said . . . that I might get F Company, but I told him otherwise. I'm holding out for 1st Battalion or nothing." He had no intention of ever working for Gordon.[210] On July 27 he wrote, "Sometimes it makes me so damned mad I could catch a boat and leave. Then I think of the madhouse up there and I think how lucky I am to be down here." He had just received a telegram from Whipple summoning him to HQ, and he feared that the colonel planned to assign him to a command in 2nd Battalion.[211] But fate intervened.

In a July 31 letter to Abbie, Parsons wrote, "Well this afternoon he [Gordon] went out on the road. . . . They turned the recon car over 7 times. Gordon is in the hospital. . . . Everyone expects him to claim internal injuries and stay in the hospital forever." Given that circumstance,

Parsons didn't object when "Old Grandma" gave him Company F: "F Company is in a hole, if you ask me, but it's going to come out and do it fast."[212] On July 31, Sgt. James Heard and his ten-man squad, part of Company F, went to work for Cpt. Walter Parsons.

CHAPTER ELEVEN
Mentasta Pass

If Major Gordon and his officers wasted the efforts of half of "Old Grandma's" regiment through July, at least the other half, his 1st Battalion, entered July making road. But if Whipple had no leadership skills, he also had no luck. In early July the soldiers of 1st Battalion confronted Mentasta Pass, and their progress slowed to a crawl. On their maps and in their imaginations, generals Sturdevant and Hoge had seen a route from Slana to the Tanana River—the old Abercrombie Trail.[213] They blithely dispatched the Black soldiers of the 97th to Slana and directed them to follow the old trail north. They would, of course, need to upgrade the trail a bit. In the generals' ridiculous plan, this stands out as the single most ludicrous element. Building his original trail north to the Tanana River, Abercrombie had crossed the Continental Divide through a notch in the Mentasta Mountains, and over the last twenty miles from Slana to Mentasta Pass he had confronted a major problem.[214]

Water—rain, snow or melting ice—from the south side of the Continental Divide flows down through the mountains to the North Pacific. In the bitter Alaska cold, the water freezes. Over centuries of geologic time, freezing water forms into glaciers. Glaciers sprawl all along the south face of the mountains in Alaska, and they slide, ever so slowly, down the face, grinding and abrading the rock beneath them into fine particles that permeate their ice. Every spring some of the ice on the front of the glacier melts, and a massive flood cascades down, carrying particles of ground rock. As the water slows at the base of the glacier, the particles fall out and, over the centuries, accumulate into a terminal moraine, a broad fan of debris sprawling out from the base of the glacier. (See Map 6 in Chapter 8, on page 69.)

From Slana up to Mentasta Pass, Abercrombie had carved a ledge for his pack trail up and along the side of cliffs made not of rock but of crumbling terminal moraine. His ledges tended to disappear under debris sliding down from above. And, more important, the outer edges of his ledges, the ones suspended in thin air, tended to crumble and fall away.[215] Abercrombie had carefully carved his narrow ledges with hand tools. His precarious trail served careful men leading nervous pack horses. But his trail crumbled and disappeared almost as fast as he could carve it. In 1942 the generals dispatched the soldiers of the 97th to find Abercrombie's trail and recreate it. To do that the soldiers would have to drive bulldozers out onto the unstable debris and carve from it a trail that could pass heavy trucks. As Twichell writes, "While working on the precipitous terminal moraine . . . the lead bulldozers repeatedly slipped off the narrow trail and 'threw a track.'" Reinstalling a tread back onto a dozer's drive sprocket, relatively routine on flat ground, became something very different in Mentasta Pass: "Doing it on a 23-ton machine that was teetering on the edge of a crumbling slope of glacial debris called for great skill and calm nerves. Eventually the 97th's inexperienced operators became masters at such on-the-spot repairs."[216]

Sgt. Lee Young came to the 97th from Engelhard, North Carolina. Trained by Caterpillar trainers at Eglin to operate a bulldozer, Lee trained others: "I was very proud when I was promoted to train the guys. I was one of the oldest operators there and I was only twenty-two."[217] Lee remembered the mountains: "We did more work in the mountains than any regiment. We followed the mountains around. The son of a gun got to know how to drop that blade to keep from tumbling down the mountain. The Army don't tell you how to do it. They just tell you you've got to do it."[218]

Sgt. Clifton Monk, also from North Carolina, had seen snow-covered Valdez as "Hell on earth" back in April. In July he operated a bulldozer through Mentasta Pass. Remembering his fellow operators, he said, "They learned real fast. If anybody tells you a colored soldier ain't a smart man and can't learn anything, you just tell them they are a dog gone liar. Those men took that machinery and built those roads (see Photo 3)."[219]

As the soldiers struggled through Mentasta, Duesenberg had men in place at Gulkana and equipment on the way. Warren Duesenberg traveled north on a scouting trip. He got only as far as the "Slana Negro Camp."[220] The army wouldn't let him go farther "because of the weather and the condition of the road." Two days later he tried again, but they still couldn't let him through.[221] The soldiers of the 97th, of course, worked up there, danger be damned. At Gulkana, Lytle and Green worked out details and dispatched subcontractors. The group of graders and bridge builders led by Osterman assembled enough men and equipment by mid-month to begin building the highway south from Big Delta. At the end of the month, the soldiers of the 97th had finally worked through Mentasta. The contractors led by Duesenberg moved up behind them to Slana on July 29. Milton Duesenberg wrote in his diary, "Set up tents for sleeping and cooked sandwiches in the open. . . . Col. Whipple payed a mid-night visit. Bulldozers arrived late. This is life in the raw." The civilians named their camp the "What Fir" camp.[222]

For the first time, the Black soldiers and the white civilians came into contact with each other. On the first morning, a Black soldier struck up a conversation with "Bubbles" Smith. When Bubbles told him he liked Alaska, the soldier responded, "I've been here 94 days, 3 hrs and 10 minutes." Bubbles assumed the soldier meant he'd been there long enough.[223] The Black soldiers teased one another and sang while they worked. And some of them teased the white men. In a letter Max Smith told his wife about passing a group of Black soldiers in the woods. One of the soldiers noticed Max's double-bitted axe. "Man o man, two blades, that man cut coming and going, boys. One blade is enough for me."[224]

Continuing problems with getting groceries to the civilians meant they lived, day after day, on corned beef. One Black soldier grinned and said, "What you boys eating, some more of that corned beef?"[225] Don Garlock remembered "a pretty good-natured bunch who seemed to like to stop and visit." The soldiers also helped out with the corned beef problem. Don wrote, "Yesterday when we were working some soldiers gave us 6 . . . cans of chili. We fixed 3 cans of it out on the job and didn't even bother to drive into camp for dinner. There were only 12 of us and we ate until we were nearly sick. Tasted better than anything we've had for weeks."[226]

The racist generals had put their plan together back in March assuming the Black soldiers would, at most, bring themselves and a willingness to follow orders to Alaska. The civilians would bring expertise and competence. It had taken too long to get the civilians in place. The Black soldiers had proved to be perfectly competent and had learned a great deal on the job. By the end of July, the 97th had well and truly taken the lead, and the civilian contractors had learned from their experience what had to be done and how to do it.

White civilian Philip Philippe wrote to his mother: "There was never a road here before, only a trail. A Negro regiment is ahead of us clearing the way for us." Camped between two engineer camps, the contractors "would be able to see firsthand the problems encountered by the engineers in preparing the base for the grade."[227] The soldiers taught the civilians how to deal with muskeg, the thick, decaying vegetation resting on bog water. They also taught them to leave the insulating moss covering and how to install corduroy.[228]

CHAPTER TWELVE
Change of Command

The trouble that brewed for "Old Grandma" came to a dramatic head on the first of August. Rumors about his personality and lack of leadership had swirled down at headquarters in Whitehorse. More important, the 97th hadn't built nearly enough road to satisfy General Hoge. At the beginning of August, General Hoge abruptly fired Whipple and turned the regiment over to Whipple's adjutant, Lt. Col. Lionel Robinson.[229] Hoge dispatched Maj. Clement Waite from his Whitehorse staff to the 97th, and Robinson put him in command of 2nd Battalion.[230] Hoge didn't get rid of Maj. Saul Gordon, but he stashed Whipple's problem child out of the way in a desk job.[231] Cpt. Charles Mitchim, who had commanded the successful 1st Battalion, continued to command it and became Maj. Charles Mitchim.[232]

Colonel Robinson came to the army from the Florida National Guard, so he had military training and experience. But he had run his family's successful construction company for years and had far more extensive experience as a contractor.[233] He had watched the bickering, incompetent officers of 2nd Battalion ruin its effectiveness. But he didn't blame the young Black soldiers. Unlike "Old Grandma," Robinson knew exactly what to do to fix it. On July 31, 1942, Robinson penned a memo. The subject was "OUR JOB."[234] He addressed it to "All Officers" and opened with "We are at WAR! We have been assigned a war mission in the combat zone! Do we all realize this very important fact?"

I would like to assemble all the officers of our regiment and discuss our mission, and ways and means of bringing this mission to a successful conclusion. But we are too scattered, and time is so short that I am forced to use this method of conveying my thoughts to all of you. . . . We are constantly making mistakes, many of them grievous errors and they must cease. . . .

We have been actively engaged in road construction since June 7th. . . . Fifty days have elapsed, and we have completed an estimated thirty miles of road. . . . We have a minimum of one hundred and ninety-five miles remaining to be done. . . . It will be some time in December when we finish. . . We are told to expect zero [bad] weather in August and most anything thereafter. It is suicidal to try to work in this country after October 30.

We are prone to complain about the hardships we encounter. What are we going to do when it gets ten-twenty-thirty and forty degrees below zero? . . . We'll build our road and we'll finish very near the end of October.

But there's no use kidding ourselves. If we don't start here and now, we'll never do it. . . . If we continue to break up equipment, we'll never do it. . . . If we continue committing the grievous errors we've been committing, we'll never do it. . . .

I'm going to list some of the things I run into every day that are keeping us from winning this fight.

Alibis: . . . In some instances, I've seen it take longer to figure out some reason for not doing a thing than it would have taken to do it in the first place. Let's not have any more alibis. I for one do not wish to hear another until after October 30th.

Procrastination: . . . Time after time it takes from two days to two weeks to make simple corrections or get a new idea started. Some officers work enthusiastically on the items they wish to do but put the others off indefinitely. All orders must be carried to conclusion at the earliest possible moment. . . .

Poor Judgment: I have seen so much of our equipment idle because of poor judgment. I have seen Carry All's [LeTorneau Scraper, known as a carryall)], D-8's [Caterpillar's massive bulldozer], D-4's [smaller version of the D-8], Power Graders [1942 versions of road graders we see at work today] . . . , and trucks, stuck hard and fast, time after time, day after day, when the exercise of sound judgment would easily have prevented it. . . . I have seen equipment placed on the deadline and remain there for weeks when the exercise of good judgment would have prevented it.

Poor Supervision: In many, many instances, I have seen equipment worth several thousands of dollars working with no officer within miles. . . . Many times, loafing on the job. Many times, I ask the enlisted men what they were doing, and they didn't know, not even why they were there! Many times, I will find two-three-four and five officers congregated where they were not needed at all. Day after day I see men in groups of fifty to seventy-five with no officers present. . . . Only two or three shovels moving out of fifty, only two or three axes moving out of dozens, only one or two pieces of corduroy moving out of the woods onto the road when the forest is alive with men. I know of at least three instances where Carry All's have been moving poor earth onto the road where gravel was required. The earth being moved was no better than that already on the road. I've seen an officer draped over the steering wheel of a pickup, sound asleep while he was supervising the work of several trucks, two tractors, and many men. What an incentive and an example for the enlisted men under him. Is it any wonder they too were loafing?

Poor Workmanship: Perhaps the most important item involved in construction of gravel roads is drainage. It does not require an engineer's degree to know that water ruins any dirt road. And yet time after time I find deep ruts left day after day to catch all the rain and saturate the roadway, soaking up water like a sponge. Many times, I find sections where graders have opened up ditches on each side of the road but in so doing, they leave a ridge of earth twelve to eighteen inches high piled up between the ditch and the roadway. This makes a perfect dam, converting the road into a great trough, perfectly constructed to catch and hold all the water possible. . . .

Don't get the idea that I think everything I see is wrong. Far from it. I see many instances of devotion to duty on the part of many officers every day. Some officers are driving themselves at a rate I do not think they can keep up. . . . Many are always on the job playing the game with all they have got.

The point I am trying to make is that all of us must be on our toes, mentally and physically, all of the time. . . . I am asking every one of you to entertain no idea of failure but to pledge yourselves to renewed effort.[235]

Robinson immediately reorganized the regiment. Two companies from 2nd Battalion would work in support, running a sawmill, delivering material, and performing other duties. He moved Captain Parsons's Company F to 1st Battalion.[236] In August four companies would build the road.

Just five days after assuming command of Company F, taking in the rapid-fire changes, Captain Parsons brought Abbie up to date:

> Things have been happening so fast day and night until I don't know what I've told you and what I didn't. Anyway the latest and "mostest" important thing to date is that Col. Whipple got his orders to report to Whitehorse at once and Robinson was to take charge. Boy was I glad to hear that news and [so] was everyone else with maybe one or two exceptions.
>
> Another thing F Co (guess I told you who is C.O. [commanding officer]) there now has been attached to the 1st Batt [Battalion] for a while. The 1st have been breaking the trail and the 2nd bringing up the rear. Getting a Co [company] that was just about on the rocks, change it into another Batt, moving from the rear well up in front, Gordon in Hospital, Whipple going out. . . . In addition, it's been raining for three days and the Gen [General Hoge] due today which is worse than rain.[237]

Actively supervising officers changed the lives of the young Black soldiers on the road profoundly. Under Whipple, their officers had sent them out on the road and then ignored them. Working for white men, these Black soldiers responded as they would have back home in the Carolinas and Georgia: they tried their best to stay out of sight and out of mind. Having to guess what the white men required of them, knowing they would get the blame if they guessed wrong, they had done as little as possible. Under Robinson, however, all of that changed abruptly. Robinson's officers joined them in the wilderness, made decisions, and called the shots. Sergeants like James Heard knew exactly what Captain Parsons expected hour by hour and day by day.

Inevitably, though, a few of the young Black soldiers didn't get the message right away. But in Robinson's regiment, actively supervising officers enforced strict discipline and adjusted the Black soldiers' attitudes, when necessary, quickly and effectively. On August 18 Parsons wrote to Abbie about three young privates who resisted doing some extra duty he had assigned them: "After three days in arrest, under guard and only two cold biscuits and water [for] a meal[,] they sent for me and said they would like to come to an understanding. Well we had an understanding, let the guard go, took down the stockade. . . . The three men are doing the extra duty, and everything is ok."[238] When a soldier edged past recalcitrance and into outright misbehavior, sergeants didn't need officers to tell them how to respond. DeWitt C. Howell commanded Company E in August and remembered "when the 1st Sergeant threw the head cook headfirst into a fire barrel to sober him up after he had become drunk from drinking all the vanilla extract in camp."[239]

Robinson's officers dealt with a few just plain bad apples. Parsons wrote to Abbie on August 22, saying, "One of my bad boys got into more trouble today! He cut a Sgt. four times. Two big cuts across the back of his head just above his neck went to the bone. The man is under guard now. I'm taking him to HQ tomorrow and expect him to get plenty and soon."[240] At least one young Black soldier broke under pressure. On August 5 Captain Parsons wrote, "To tie it all, one of Joe's men shot another one of his men and he is not expected to live."[241]

Lt. Joseph Raso, who had led a pack mule to Mentasta Lake in June, had assumed command of Company B in July. When they had a little extra time, his Black soldiers gambled in their tents and shot craps. One soldier, Pvt. Joseph Anderson, lost and continued to lose until one day he lost his temper, picked up a rifle, and shot Pvt. Charles Vance in the stomach.[242] Vance died at the Ladd Field Hospital in Fairbanks a few days later. A month after that, Anderson hanged himself in his cell.[243]

But in August the work progressed, everybody pitched in, and morale improved. Captain Parsons's mood reflected that of his men. On August 10 he wrote to Abbie, telling her that he was doing what he had wanted to do for months—to "build this damn highway. Now I can hurry it along and get back home. They still say us colored folks are going SOUTH this winter. That's ok with me in a big way."[244] Through August, with the specter of an Alaska winter looming—days still warm but nights getting colder—the men of the 97th accelerated away from Mentasta Pass. Soldiers and dozers surged north across Station Creek, past Mineral Lake, to the Little Tok River and across Clearwater Creek, headed for the Tanana River and their portion of the Alaska Highway.[245]

CHAPTER THIRTEEN
Progress

As they raced north in August, Robinson's newly energized and efficient regiment continued to face obstacles. But it dealt with them far more effectively than it had before. Getting supplies, especially endless drums of fuel, remained the single biggest problem. A civilian contractor had moved in behind them to bring supplies out from Valdez to Slana. But the soldiers moved farther away from Slana every day, and between them and Slana lay the Continental Divide and the streams and rivers of the Yukon River system. Heath Twichell, in his *Northwest Epic*, described one creative response. The soldiers of the 97th built rough sleds on runners of logs, and their dozers dragged the sleds carrying supplies and fuel as they moved through the Tanana Valley (see photo 5). Of course, they had to replenish the fuel and supplies on the sleds. To do that they used the Little Tok River to its mouth on the Big Tok River. Twichell writes, "Enough supplies for several additional weeks came floating down the Little Tok behind them, to be caught by a log boom after bobbing and bumping along over rapids and sandbars for 15 miles: fuel in half-filled drums, rations, spare parts, and miscellaneous items in lightly-loaded pontoons."[246]

Duesenberg's civilians watched. "At one point in coming down the Tok Valley," Duesenberg wrote, "one 97th crew had floated pontoons of supplies, rations, and half-filled gas and diesel fuel barrels down the Little Tok River fifteen miles to where they were caught in a log boom."[247] In his

August 10 letter to Abbie, Parsons described how that process worked on the ground:

> We are having plenty hell. Most of our supplies are behind and the ... companies out in front ... are just getting by—packing our food in. ... We sent food down the river in boats but could not get the boats back up stream. Most of our fuel we get by half emptying full drums into empty drums and dumping both into the river. After 30 or 40 drums we send men walking down each side of the river to keep the drums on the way.[248]

Following the painted marks left by the advance survey crew, catskinners led the way through the valley, piloting their big dozers through the woods, fording small streams, and ferrying over rivers. Occasionally they had to divert from the path to pull other dozers and vehicles out of the mud, but mostly they focused on gouging out a continuous right of way by knocking down trees. Few trees could resist the big D8. When the heavy steel blade slammed into the tree, the tree would fall. Sometimes, though, as the tree suddenly jerked, falling forward, its top would whip back, break loose, and fall toward the dozer. A big treetop could crush a man, and an experienced operator could bail out into the mud in a split second.[249] In August, Willie Calhoun, a member of Sgt. James Heard's squad in Company F, failed to bail out quickly enough: "A falling tree fell on his left shoulder, left hand and left chest. He was treated at a field hospital for three weeks."[250]

If the catskinners became experts, so did the young Black soldiers who worked all around them. Some drove trucks. Nathaniel Boyd, Luther Mack, and Spencer Fisher from North Carolina and John Lee from Tennessee learned along with hundreds of others to guide their trucks along the mountainsides and through the clutching mud of Alaska.[251] The bulldozers and trucks broke—often. Other young soldiers learned to fix them. Men like Alphonso Martin, a janitor at Clemson College back in South Carolina, patched the trucks and bulldozers together and kept them working.[252]

When a catskinner got his dozer stuck or disabled, he and the soldiers working around him learned how to get a stuck or disabled dozer back in action. And as the dozers laid down trees, soldiers with hand tools scrambled over the right of way, shoveling dirt, raking the right of way clear (see photo 7). Because the dozers and the men who swarmed around them worked through Alaska, they learned to deal with muskeg. At first glance muskeg looked like dirt. It froze to solid ground in winter, but the top few feet thawed to boggy muck in spring and fall. A layer of vegetation covered and insulated the muskeg, so a few feet down the water remained frozen and the muck had a bottom—until soldiers cleared away the insulating vegetation. Then the bottom melted away. Bulldozers would move out on mud, slewing through it, the ice would melt, and then, as they passed through again, they would stop slewing and start sinking (see photo 6).[253]

The soldiers found a solution in corduroying the roads. The big cats slogged through once, knocking trees down, then the soldiers on the ground would cut thirty-foot logs and drag them to lay side by side across the right of way (see photo 8). Covered with gravel and dirt, the corduroyed construction offered a relatively smooth roadbed. But Alaska's muskeg didn't give up easy. Often the corduroyed trees simply sank down into it. When that happened, they installed another layer, and another, and kept layering until the corduroy stopped sinking.[254]

The young soldiers and catskinners also became experts at building and installing culverts. Alaska presented as rugged a terrain as exists anywhere on earth. The path of the future highway ran along mountainsides where, over millennia, water flowing down from higher elevations had coalesced into streams that had carved deep channels across the path. The catskinners and their bulldozers could fill the channel to make the road relatively level, but then the fill would dam the stream. Water would simply accumulate behind the dam until it washed out the road.[255] The soldiers built rough, square tunnels of logs and timbers for the stream to flow through—in other words, culverts (see photo 9). On top of the culverts, the bulldozers could pack and fill as much as necessary to level the road.[256]

Culverts dealt with the small streams, but for larger streams and rivers the soldiers had to build bridges. They had no concrete or steel, of course, so they built bridges of timber, driving pilings every twenty feet or so and spanning them with timber stringers. Across the stringers they laid planks, or decking, for the roadbed (see photo 10). The soldiers moved a sawmill through the woods behind the road builders, turning Alaska's bigger trees into pilings, stringers, and decking.[257] Civilians built the big bridges, but the young Black soldiers encountered and bridged creeks and small rivers all along the way from Carlson Creek to the Little Tok River.[258]

Finally they all had to eat. Men like Thad Bryson from North Carolina learned to cook at Camp Blanding and Eglin Field. In Alaska they learned to do it in tents using a dangerous white gas cookstove.[259]

Major Mitchim, Lieutenant Gordy, Lieutenant Clark, and seven soldiers of the attached 29th Topographic Unit worked out ahead of the 97th Engineering Regiment, laying out and staking the path. The four lead companies built road behind them. By August 10 Mitchim's party had worked sixty miles out from Slana and just ten miles from the Tanana River. Company A worked one mile back.[260] Ten days later, Robinson's progress report for August 20 referred to Mitchim's crew as being across the Tanana River. They had crossed the river at the mouth of the Big Tok River and turned to stake along the north bank toward Canada. The soldiers of Company A camped just four miles from the river at the edge of what the soldiers remembered as the Tanana Flats.[261]

The Tanana River expands to a width of several miles during the spring thaw and then shrinks dramatically during the rest of the year, leaving a broad, flat, dry area on either side of the flowing water. In August Company A could easily cross the few miles of dry riverbed to the river. The Company A commander, Cpt. Andrew McMeekin, had left the company bivouac on August 15 with a transit, following the most direct tangent to the river. A single D8 had followed him, and that night a few soldiers of Company A camped on the bank of the river.[262] In Northwest Epic, Heath Twichell sums it up this way: "By mid-August the trail from Slana . . . to the Tanana River was finally open, the last miles pushed across the river's floodplain by Company A in a single day."[263]

At the junction of the Tok and Tanana Rivers, the place that would be known to the soldiers as Robinson's Crossing, Company A had a problem. Pontoon boats and bridging material, ordered well in advance, had come to Skagway instead of Valdez and wound up in Whitehorse, Yukon.[264] Rerouted, the equipment headed up the Gulf of Alaska, but company commanders in Robinson's regiment knew not to wait around for someone else to solve the problem. A sternwheel steamboat, traveling from Fairbanks to trade with the Tanacross, Tetlin, and Northway First Nations, came down the river. McMeekin hailed it and quickly made a deal with its owner/captain. The old steamer ferried Company A and at least some of its equipment across. And the soldiers of Company A turned to follow Mitchim's crew toward Canada (see photo 11).

Three more companies, including Captain Parsons's Company F, arrived at the river right behind Company A. At the end of the month, when he had time to write, Parsons brought Abbie up to date: "The Tanana River has given us some trouble since it's too wide for our pontoon bridge. We have had to build a ferry and everything but our D-8 tractors is going across on the ferry."[266] By the time Parsons wrote that letter, a company of the 73rd Pontoon Engineers had made it to the river.[267] The soldiers of the 73rd assembled the pontoons into a ferry and powered it with four outboard motors. Their pontoon ferry could haul heavy equipment across.[268] Robinson's progress report for August 31 locates five companies on the north bank of the Tanana, ready to build the road toward Canada.

In August the civilian contractors still struggled with supply and organizational issues. Nevertheless, they started working. General Sturdevant's and General Hoge's expectations that civilians would do the real job had gone by the board. The contractors followed competent soldiers and who could not have made the progress they made without the contractors' skilled and dedicated support.[269] Behind the soldiers a contractor hauled

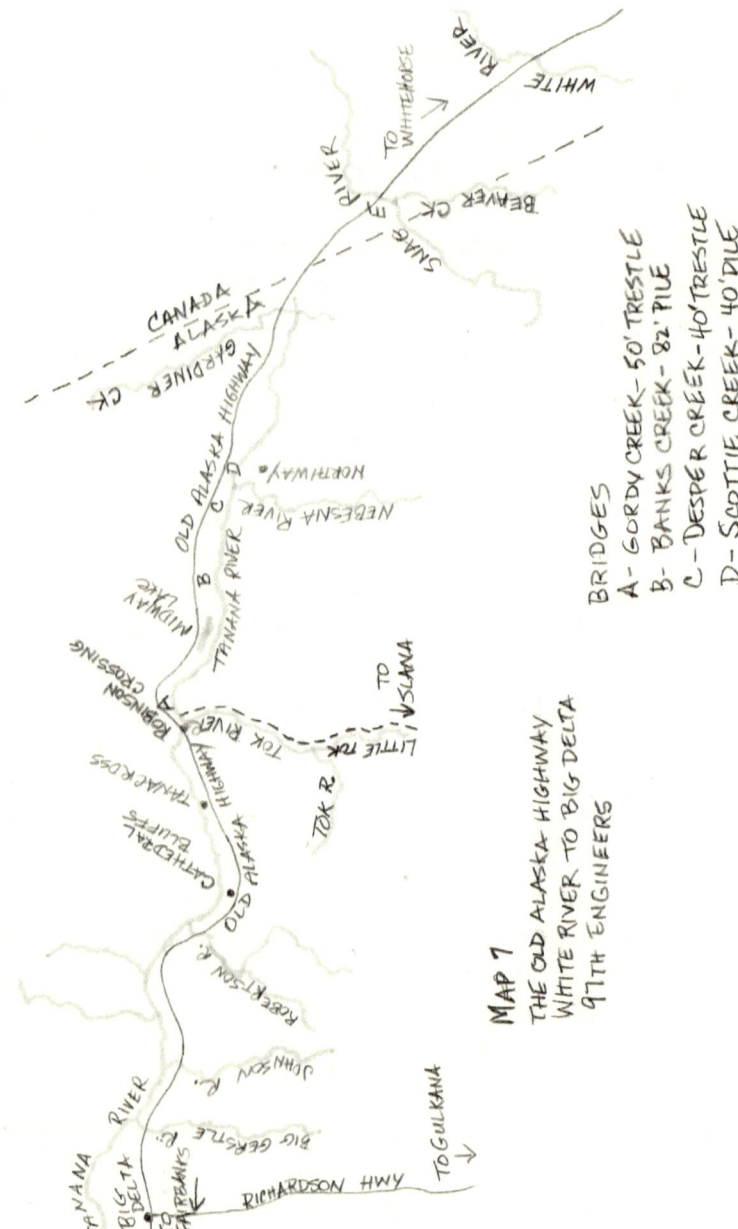

MAP 7   The Old Alaska Highway, White River to Big Delta

supplies to Slana. Another contractor worked to upgrade the road past Slana. Duesenberg's crew moved on August 8 through Mentasta Pass and out to a camp they named "Thunder Mountain" near Mineral Lake. Crews out of Thunder Mountain took their turn at the road from Mentasta Lake to Mentasta Pass, widening and surfacing the fragile roadbed. Other crews followed the soldiers toward the Tanana, maintaining and upgrading the road they built.[270]

At the Tanana River, bridge contractor Weldon Brothers hurried to construct a 365-foot temporary bridge over Robinson's Crossing. They cut trees, built a rough log derrick, and set up a gasoline engine and winch to drive pilings. Between the pilings they installed timbers salvaged from an old collapsed bridge at nearby Chitina on the Copper River.[271] Contractors led by Osterman worked south out of Big Delta through August. Alaska introduced them to muskeg, and they adopted the army's corduroy solution. They also encountered two large, glacial rivers—the Big Gerstle River and the Johnson River—one thousand to two thousand feet wide with many individual channels. Heavy machinery could ford them, but bridge contractors hurried to construct temporary wood trestle bridges so that trucks could cross.[272]

The water level in the rivers fluctuated wildly. After warm days and heavy rains, the streams turned into raging torrents, carrying gravel and "immense boulders."[273] The bridge builders built with native timber and set up sawmills to carve it out of the trees they cut. And they traveled down to Chitina to salvage timbers from the same collapsed bridge Weldon Brothers had plundered for the Robinson's Crossing bridge.[274]

On the fly in August, Lytle and Green changed the plan for using Duesenberg's crews. Across the Tanana the crews would split. Warren Duesenberg would lead half of the men across the Tanana and then work north along the river toward the Osterman crews working down from Big Delta. The two crews would meet at the Robertson River. Milton Duesenberg would lead the other half of the men across the Tanana and follow the Black soldiers south toward Canada.[276] As the soldiers of the 97th and the civilian contractors who followed them worked south, the soldiers of the 18th

Combat Engineering Regiment worked north through Yukon Territory toward Alaska. The climax, the end of the project, would occur when the soldiers of the 18th met the soldiers of the 97th.

Col. Earl G. Paules brought the 18th up the inside passage to Skagway in early April. They were the second regiment to arrive on the project and the first in Yukon. The white soldiers of the 18th rode the narrow-gauge White Pass and Yukon Territory railroad up to Whitehorse and started building north from there. As generals Sturdevant and Hoge had confidently expected, the well-organized, experienced combat engineers performed magnificently, turning out mile after mile of Alaska Highway.[277] The soldiers started with the stretch of highway from Whitehorse to the Slims River at the south end of Kluane Lake, a distance of 148 miles. In his official history of the 18th, Fred Rust describes building that stretch: "We were lucky in the conventional character of the work in this curtain raiser. It was like building a road in any rolling, wooded country veined with streams, and there were just enough trouble spots to broaden our experience without overwhelming our innocence."[278] The soldiers of the 18th had built bridges. And they struggled with a few other problems along the way, but "on the whole it was ordinary roadbuilding."[279]

At the Slims River and Kluane, things changed in July: "Kluane, beautiful and indifferent squarely blocked our path."[280] The soldiers had to build 8 miles of road across the foot of the lake to Slims River, bridge the river, and then build 40 miles of road north along the lake to the Big Duke River. Still, "We had good luck in meeting the difficulties of this sector and pioneered it in six weeks."[281] By mid-August the white soldiers of the 18th had reached the Big Duke River. They had about 160 miles of road to build to reach their goal: the Alaska border. The Black soldiers of the 97th worked to get across the Tanana, 108 miles from that border.

Generals Sturdevant and Hoge had brought Paules and his 18th north with full confidence in their ability to do the job. Month after month the 18th had built more miles of road than any other regiment on the project, justifying the generals' confidence. The generals had brought the 97th north with a very different set of expectations. Now the race to the border was on, pitting these two regiments against each other.

CHAPTER FOURTEEN
Reorganization

As August came to an end, the soldiers of the 97th reached and crossed the Tanana River. Major Mitchim's survey party had already moved south, driving stakes to locate the regiment's portion of the Alaska Highway. At the beginning of September, Captain McMeekin's Company A bulldozers followed Mitchim's stakes, pushing aside trees and other obstacles, gouging a rough path through the woods.[282] But as the nights cooled and the leaves changed color, Colonel Robinson knew he had a big problem. One hundred and eight tough miles remained between the regiment and the international border, and to get there by October 30 they had to move faster.[283] Through August, as companies B, C, and F had followed Company A through the Tanana Valley, turning their path into a rough road, their organization in separate companies had slowed their progress. Captain Parsons remembered the problem: "All the Companies . . . were equipped the same, had assigned sections of the road and were 'leap frogging' each other to get to their second, third and subsequent assignments. This method led to utter confusion. When a company finished its assigned sector, it had to pass through the unfinished sector of all other companies to get to its new assignment."[284] On August 31, Colonel Robinson fixed the problem. He completely reorganized the three companies. Parsons described the colonel's fix to Abbie in his letter that night:

> The Colonel called several of us in to Hq [headquarters] tonight and is reorganizing the 1st Battalion which as I have told you now consists of four companies. I'm to

give up all my heavy equipment to Joe and Jim [Lt. Joseph Raso commanded Company B and Lt. James Coleman commanded Company C] and in turn they give me all the men in their companies less what it takes to run the equipment. I'll have about 500 men when the change is made and will build all the bridges, culverts . . . etc.[285]

Captain Parsons reported to Abbie on September 3: "Tomorrow Coleman, Raso and myself all move [together] . . . and will camp together from now on under the new setup I told you about."[286]

In September, working south behind Company A, Lieutenant Raso's Company B and Lieutenant Coleman's Company C would not work on assigned sections; instead they would move from problem to problem, from culvert to corduroy, to bridge and back again. Captain Parsons's enlarged Company F would take care of kitchens, food, spare parts, and fuel. And when Coleman and Raso needed men on the ground, Parsons would provide and figure out how to transport them.[287] Parsons struggled at first, especially with transportation. On September 8 he messaged Lt. Aurelio Basile at headquarters: "Trucks needed now more than ever before. . . . We have two Chev [Chevrolets] on D.L. [deadline] but can get some off if parts are received. Am sending [2nd Lieutenant] Dudley in with list . . . give him a break."[288] That same day, Parsons also messaged Colonel Robinson: "How many and when can we expect trucks to move our camp? One platoon moved yesterday, another today. Balance of Company should move this P.M. or tomorrow at latest."[289] On September 10 Parsons generated a flood of messages to headquarters. To Robinson: "No trucks have arrived here to date. Have only two 1½ ton and work is eighteen miles from here. Must have additional trucks at once if [we] expect to ever move forward."[290] To a Captain Sprigg at headquarters: "Yes another 1½ ton on D/L but should be off by noon as we are working on it now. Two 1½ ton have reported so far to us and one or two to C which I am using. Have been hitch hiking men on oil trucks all morning."[291]

Parsons struggled to feed his five hundred soldiers along with the catskinners and mechanics of companies B and C. His cooks used white gas stoves. To Robinson he wrote, "White gas as ordered by Waite two days ago not received. Our fourth kitchen going out this AM." And, of course, cooks must have something to cook. "Have three days rations."[292] On September 16 he wrote to Major Waite: "3 drums of white gas needed for cooking at once." And on September 23 he wrote to Lieutenant McMillian at H&S: "We are out of potatoes. Have received no flour, lard and sugar for the last ten days."[293]

Finally his five hundred soldiers had critical work to do. On September 13 Parsons messaged "Sprigg or Mrozek . . . Twenty culverts to be built in 10 miles south of here. Am sending truck. . . . See that large quantity of large nails and drift pins gets on this truck."[294] He juggled his men, moving them to where Coleman and Raso needed them, working through his subordinate platoon leaders. On September 13, for the moment with Company B, he dispatched a messenger to Lt. George Hill and Lt. Leigh Robinson telling them that he was sending trucks to "move men from Hill to Robinson to work south." He instructed them to "save [Sergeant] Calloway's men to last as they are to move . . . along with Hill. . . . Have wired Raso and Coleman to send chow truck to Robinson. If possible, use it for you both."[295]

Where the catskinners of Company A had cleared over rough terrain, descending steeply to a creek and then climbing back up, the soldiers of Company F built culverts to pass the water, and the catskinners of Company B or Company F pushed dirt on top of them to fill the dip. Where Company A had cleared over mud and muskeg, Parsons's soldiers cut logs, Raso or Coleman's bulldozers dragged them to the road, the soldiers on the ground muscled them into place, and the dozers pushed fill dirt to cover them. Where Company A had forded streams and rivers, soldiers assembled bridge timber and decking; bulldozers and graders dug the approaches at either end.[296]

Colonel Robinson's reorganization worked. The companies raced through the woods, and Duesenberg's civilians raced right behind them, grading, graveling, and upgrading some of the bridges. Together the soldiers and civilians completed forty miles of pioneer road in the second ten days of September—an unbelievable four miles a day.[297] The junction of the Tok and Tanana Rivers served as the hub of everyone's effort in Alaska through the fall. The soldiers building the road and half of Duesenberg's crews worked south from there. The rest of Duesenberg's crews worked away in the opposite direction toward Big Delta and Osterman's crews, who were working south to meet them. A joint military/civilian headquarters mushroomed out of the woods and acquired several different unofficial names, including Robinson's Crossing and Tanana Crossing. The contractors took to calling it Tok.[298] The new headquarters acquired offices, supply dumps, and motor pool and repair facilities. The officers of Robinson's H&S Company settled into offices in temporary buildings or tents. Their soldiers fixed broken trucks and heavy equipment. The officers ordered food, fuel, vehicles, and repair parts, and the soldiers of Company E unloaded, sorted, stacked, and then reloaded all of that to send out to companies and crews on the road.

Robinson turned some of the soldiers of Company E into carpenters. In September they built buildings for the new headquarters.[299] The carpenters of Company E and the culvert and bridge builders out on the road used enormous quantities of lumber. The soldiers of Company D had operated a sawmill back near Lake Mentasta since the beginning of August, and Robinson left them there for the time being.[300] Contractors hauled fuel and supplies 250 miles up from the Valdez docks to Slana and then over the new road to the Tanana River. The contractors called the road from Gulkana out to the Tanana River the "Tok Cutoff."[301] The heavy traffic ravaged the roads, and contractors based in Valdez, Slana, and the new headquarters worked to repair them.[302]

Men working at headquarters, both military and civilian, slept and ate in large clusters of tents and temporary buildings. They didn't have to move, so they settled into their accommodations. Out on the road, soldiers lived in bivouacs and kept them as close to their work as possible. Parsons wrote to Abbie on August 22: "The balance of my Co [Company] just moved in today from our last camp about 25 miles back and we expect to move again in a few days. We move so fast we can hardly keep up with ourselves."[303] As organization by company went away, company bivouacs went too. The soldiers still slept on the ground in crowded twelve-man tents. Cooks still prepared food in kitchen tents (see Photo 13). Bivouacs still simply squatted along the side of the road adjacent to the work.[304] But as Company B and Company C scattered to problem areas along the road, Company F soldiers and kitchens scattered with them. Bivouacs housed an ever changing cast of soldiers from different companies.

In September Alaska weather made living under canvas considerably more difficult. In a letter to Abbie, Parsons wrote, "We had to heat our water this morning and melt the ice before we could wash our faces."[305] Hoge's progress report for September noted heavy snow in Alaska on September 15, which remained on the ground until September 21.[306]

Every two or three or maybe four days, at the end of a hard day's work, the soldiers packed their personal gear and took down the sleeping tents. Each man had two duffel bags, and most carried "private boxes" with family photos, letters, and other personal treasures. The men loaded personal gear, tents, and equipment in the back of deuce-and-a-half trucks and then climbed up and packed themselves on benches under the canvas cover. The cooks and kitchen police (KP) struck and loaded the kitchen tent, and the convoy rumbled over the miles to the next bivouac. A detail dug a latrine. The tents went up again and personal gear got stashed.[307] After a few precious hours of sleep, the soldiers emerged from their tents and returned to their battle with the road.

Near the end of September, past Gardiner Creek, the soldiers ran into continuous waist-deep muskeg. Trucks couldn't haul the men or their camp through muskeg, so the soldiers resurrected a tactic they'd developed back in the Tanana Valley: they built sleds out of spruce logs, loaded the sleds instead of trucks, and towed them through the muck with bulldozers.[309] The soldiers became as filthy as their road. Their tents stank because the soldiers stank. In September lakes and rivers offered only frigid water that no one could bathe in. Captain Parsons wrote to Abbie that he had recently seen an officer friend he hadn't seen for a while: "Did he look a wreck. He didn't look like a soldier but more like a tramp. As a matter of fact, most of us do look like tramps."[310]

The soldiers managed to wedge a little frivolity into the schedule. Back in August, Parsons had bought his son Walter a Husky puppy and named him "Tok" (see Photo 12).[311] Lt. DeWitt C. Howell remembered a company mascot, a black bear cub named "Dynamite."[312] Lt. Walter Mason remembered two bear cubs, "Dynamite" and "TNT." The soldiers had to shoot the mother bear when she attacked one of them, and they adopted her babies.[313]

The young soldiers worked incredibly hard in cold and then heat and in incessant rain. In addition to surviving the hostile weather conditions, the nature of their work subjected them to constant risk of injury. The soldiers drove vehicles with cannibalized parts, sometimes without brakes. They patched broken tools together with wire, tape, and ingenuity. They worked brutal hours swinging axes, felling trees, and piloting vehicles through mud and along steep mountainsides. The dangerous work lacerated skin and fractured bones on a regular basis. Even the cooks lived with risk. They used temperamental white gas stoves whose burners often refused to light until the cook literally poured white gas on the burner and then struck his match.[314]

The army had attached a medical unit to the 97th, and aid men followed the soldiers through the woods. The aid men served as first responders, sending patients with serious illness or injury to battalion aid stations. The aid stations sent the most serious cases to a field hospital that traveled with the Headquarters and Service Company. The most serious went to hospitals in Valdez or Fairbanks. Doctors and dentists visited the companies sporadically, holding sick call and fixing teeth.[315]

The soldiers worked through the woods and muskeg as fast as they could. Duesenberg's and Osterman's civilians worked more slowly, upgrading the road to a higher standard. They moved fast but not as fast as the soldiers. Back in August, Osterman's civilians had moved into portable wooden barracks built on skids that could be towed along as they progressed. Clarence Kallsen wrote home about the barracks, constructed of rough wood that was cut and milled right there on the road: "If one of these shacks was setting in the country at home some farmer would probably drag it into a pasture for his hogs.... But to us they're just like palaces after living in the tents so long."[316] Kallsen described the setting in his barracks: "I sit here at the table writing this letter. At the right the men are turning in for the night. One is setting the alarm clock, another is taking off some smelly socks, and one just came in making a lot of noise. The stove is right behind me with the lamp hanging above it. Another man is just cussing the stove as he burned himself on the door, in other words, one happy family."[317]

The Duesenbergs didn't use towable barracks. They could commute to the work site from relatively permanent encampments. When they moved a camp, they sent crews ahead to clear the campsite and set up a large tent for a mess hall and large sleeping tents for the men. By September civilians' tents had acquired wood floors and sidewalls. Their tents were equipped with generators and electrical wiring.[318] From September on, as cold weather descended, the civilians enjoyed at least some comforts. The Black soldiers, however, did not have nearly as many.

CHAPTER FIFTEEN

Civilians in Uniform and the Oncoming 18th

The army, in 1942, imposed military routines on its soldiers. Soldiers woke up together at the same time every morning, assembled in platoons and companies, and stood at attention for roll call. Officers formally inspected the men, their uniforms, their gear, and their rifles. Platoons and companies drilled and marched in formation. Enlisted men and officers exchanged salutes.[319] These routines literally defined what it meant to be a soldier, and the army had turned the young Black men of the 97th into soldiers by training them to follow those routines. In September, however, for three full companies of the 97th, Robinson's reorganization rendered those routines impossible.[320] Normal bivouacs by company went away. Soldiers from Company F bivouacked with soldiers from companies B and C. Soldiers from companies B and C ate from Company F kitchen tents. The scattered troops would not fall out with their company for roll call or inspection. They would not march or do close order drill.

The companies of the regiment kept track of their men and locations with mind-numbing morning reports, and in September the morning reports from companies B and C looked deceptively normal. Company B moved eight miles on September 3. On September 11 they moved twenty-four miles and "arrived at 1:00 am." On September 19 they moved another twenty-four miles, and on the 27th they moved thirteen more.[321] Company C reports recorded a similar pattern. But these companies moved heavy equipment and operators from problem to problem relatively easily.

111

Company F morning reports best reflected the turmoil of Robinson's reorganization. First, they recorded the influx of hundreds of soldiers from Company B and Company C. Two pages of handwritten names follow the words "Attached for rations and duty" on September 6.[322] With the hundreds of men scattered with companies B and C, Captain Parsons's clerk didn't even try to record their locations. In an undated letter from September, Captain Parsons shared the following with Abbie: "Have been so busy the past few days I've hardly had time to think. We are really moving along now and getting my large gang and their things forward and keeping the work going is quite a job. I've men and equipment at three camp sites now—expect to open another tomorrow and close another. We no sooner get unpacked at one camp before we pack and move on."[323]

Moving through woods, across muskeg, and over mountains, their rough road spooling out behind them at an ever increasing speed, the three companies merged and reorganized into a road-building team. Their officers maintained discipline and held their soldiers responsible for mistakes and errors in judgment. But the officers and soldiers of companies B, C, and F had a road to build deep in the Alaska wilderness with temperatures already falling, they had no time to waste. Cpt. Howard Garber remembered, "We did not hold any formations nor drills, and it wasn't necessary to call the roll as the men had no place to go AWOL."[324]

Robinson's reorganization changed the lives of the young Black soldiers in companies B, C, and F profoundly. Their white company commanders and platoon leaders demanded nothing but incredibly hard work and efficiency. Sergeant Heard, the men in his squad, and the rest of the Black soldiers in the three companies became, in effect, "civilians in uniform." Robinson's officers knew that one day, with the highway emergency in the rear view, the regiment would have to return to normal. They would have to reorganize into normal companies and battalions. They would have to reimpose normal military routines. The regiment would go on to other missions and it would have to do so as a normal engineering regiment. In Alaska in September, however, that didn't matter.

As the officers and men of the 97th worked south through Alaska in September, expecting to complete the road at the international border and hoping to finish and go home for Christmas, two developments, one down in Yukon and another in far-off Washington, changed everything. In Yukon

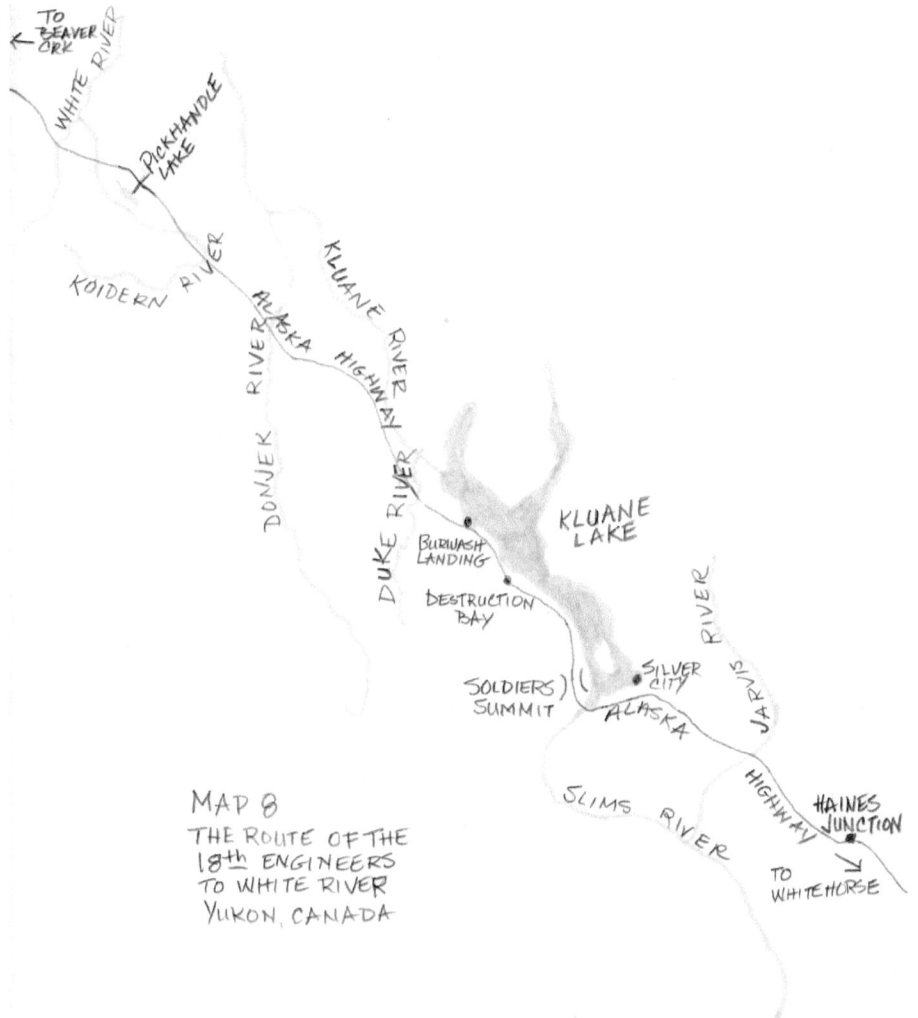

Map 8   The Route of the 18th Engineers to White River Yukon, Canada

the white soldiers of the 18th Engineers worked on their stretch from the Big Duke River to the Donjek River. Heavily wooded ground, creased by small ridges, presented no real problem, but the woods soon gave way to something very different. In his capacity as official regimental historian, Fred Rust prepared an official account he titled "Building the ALCAN Highway." Describing the sudden change in the terrain, Rust described it as looking like a "burnt over swamp."[325] But this forest looked the way it did, not because of fire but because it grew out of permafrost. The trees grew out of a layer of rotted vegetation over frozen water, deep frozen water.

For the soldiers of the 18th, the first inkling that they had a major problem came from soldiers on kitchen police. "The KP's searched far and wide for a spot to dig a large pit but no matter where they went there was frozen ground. One man on KP there for a week still complains that he had to dig "21 garbage pits."[326] The 18th had developed a road-building strategy that they had used all the way up from Whitehorse. The soldiers of Company A led the regiment, clearing a path. The other companies followed to do the rest of the work, to turn the path into a road. Suddenly, there between the Big Duke and the Donjek, they found Company A's path impossible to follow. Clearing the path exposed the permafrost to sunlight, so it melted. The path "was muddy, and our attempts to cut through merely exposed more frozen ground to thaw. Equipment bogged down everywhere."[327]

The companies that followed didn't just find Company A's path worthless as a roadway; they couldn't even use it to move ahead:

> Moving . . . a short distance beyond the passable roadhead, necessary for each company as it bit into a new section, was a nightmare of stuck trucks and broken axles. [Company] A's clearing became a deeper aisle of mud with each day's exposure to sunlight. Trucks simply took off through the woods on either side, cutting between trees or knocking the smaller ones down. The ground was so soft that one truck could not follow in another's tracks without bogging down. Each driver took his best shot and kept going as long as he could. Sometimes you would see a D-8 hauling a "train" of three or four trucks, dragging them through the gumbo.[328]

It took a while to realize just how big a problem they faced, and Company A continued to forge ahead, carving a worse than useless path all the way to and across the Donjek.[329]

Effectively, working over permafrost rendered the 18th's road-building strategy futile. Figuring out how to deal with permafrost, changing their strategy, took time. And the approach they devised slowed the regiment's progress to a crawl.[330] The men of the 97th had expected to complete their road when they met the men of the 18th at the international border. As permafrost slowed the 18th, that expectation disappeared. On September 28 Parsons told Abbie, "We should meet the 18th in about two weeks. Looks like we'll beat them to the border and in their own end of the field."[331] Parsons and his fellow soldiers remained optimistic about the more important issue of when they would complete the road and get out of Alaska. On September 22 Parsons wrote to his oldest son, telling him they "will be done in about three weeks." He hoped to celebrate Christmas at home with his family.[332]

But Gen. Brehon Somervell, at the top of the chain of command for several army corps, including the Corps of Engineers, fired General Hoge in September and organized a new Northwest Service Command in Whitehorse under Gen. James A. O'Connor.[333] He had a notion that he might want to build a road north from Fairbanks to Nome the following spring, and he intended to use the 97th for that project. Moreover, his newly installed commanders in Whitehorse planned to use the soldiers of the 97th to keep the northern section of the new Alaska Highway passable through the winter. Most of the civilian contractors would leave Alaska to return to work in the spring. The soldiers of the 97th would winter in Alaska. The Black soldiers remembered April in Valdez, and that had been only an Alaska spring. Nobody in the regiment wanted any part of an Alaska winter. But, of course, it wasn't up to them.

CHAPTER SIXTEEN

Beaver Creek

The white soldiers of the 18th had found a way to deal with permafrost in August. First and foremost, they learned not to disturb the insulating vegetation. Since bulldozers disturbed it simply by driving over it, the men resorted to clearing the land with hand tools.[334] Their progress slowed to a crawl, and the lead company didn't cross the Donjek River until August 31.[335] A floating pontoon bridge got them across the river, and most of the regiment, trying to build road north to the next river, the White, "floundered energetically for six weeks." They averaged less than a mile a day.[336] Because turbulence in the main channel of the Donjek repeatedly took the pontoon bridge out of commission, one company spent the whole month of September building a more permanent bridge across the Donjek's three channels. And as the road over permafrost inched north from the Donjek, Company E came behind, spreading gravel.[337] At least with the gravel, the road qualified as "finished" to Alaska Highway standards.

In a race to the international border with the Black soldiers of the 97th, the white soldiers of the 18th fell further behind with every passing day through September and into October. Captain Parsons, in a letter to Abbie on October 5, reported that the advance survey party would cross the border on October 6: "McMeeken [sic] should reach to boundary by tomorrow and we will follow in a few days." The lead cat arrived at the border on October 12.[338] The Black soldiers of the 97th had won the race. They paused briefly to celebrate their triumph. Parsons reported to Abbie on October 12, "We all took the day off yesterday and celebrated our reaching the boundary. It was a long, hard fight to get here, but we made it on schedule. The 18th didn't make it so now we have to go on until we meet them."[339] The Black soldiers celebrated too. Someone climbed to the top of a tree and hung a sign reading "Los Angeles City Limits." But the soldiers of the 18th, still down in Yukon, did not celebrate. They grumbled and muttered about whose section of road had been tougher. But grumbling didn't change the simple fact—Robinson's Black soldiers had got there first.[340] The white 18th Engineers' "record-setting pace over the 150 miles from Whitehorse to Kluane Lake during June and early July had established them as the highway's undisputed road-building champs."[341] But now "the highway's mileage champs had been beaten in a fair race" by the "'practically useless' [quoting General Hoge[342]] Black soldiers of the 97th Engineers."[343]

Meanwhile, on September 24, down in British Columbia, two regiments, the 35th and the 340th, had met at Contact Creek and opened the Alaska Highway from Dawson Creek to Whitehorse. "Many miles of filling and grading in both directions from Contact Creek remained to be done, but the Army knew a good public relations opportunity when it saw one. On September 22, two young soldiers . . . loaded . . . a Dodge half-ton weapons carrier and left Dawson Creek with orders to get through to Whitehorse."[344] Averaging fifteen miles per hour, it took them five days to get there. On September 27, photographers captured two grinning soldiers chatting with a Royal Canadian Mountie in front of a dusty weapons carrier that bore a freshly painted sign: "First Truck, Dawson Creek to Whitehorse."[345] The army and the Alaska Highway project had caught the attention of the press. Correspondents descended on Yukon Territory, observing and reporting the effort to "close the last gap in the Alcan Highway, a final effort to finish a race now watched by millions."[346] Race had dominated the experience of the soldiers of the 97th. Now, in October, came "a different race."[347]

The success at Contact Creek put Col. Earl G. Paules, commander of the northern sector, squarely in the hot seat. The last gap in the highway lay between his two regiments, the 97th and the 18th, and he needed them to meet—soon. With luck the 18th could complete a graveled highway as far as the White River. The 97th and the contractors graveling behind them had completed highway as far as the border. The fifty-five miles between the border and the White, though, would run over a vast lake of permafrost. Paules had one advantage—in October temperatures had dropped and frozen the permafrost to a solid base. Across that last fifty-five miles he would ignore the permafrost. He would build the highway on the temporarily solid base of ice. The press wouldn't know the difference, and when the permafrost melted in spring, civilian contractors could deal with it.[348] Paules ordered Colonel Robinson to speed up the 97th and build his road over the frozen permafrost as quickly as possible. Duesenberg's civilians, making plans to go home, got a nasty surprise: "It had become apparent ... that the 18th Engineers coming from Whitehorse would not make it to the border and that the 97th Engineers would need the civilian assistance in completing the pioneer trail to the White River."[349]

In his letter of October 7, Parsons described building over permafrost to Abbie: "It seems funny, but we are all praying for cold weather and unless it stays cold, we won't be able to finish on time. The ground along here is marsh covered with heavy moss and small trees. On warm days it's hard to walk in the water-soaked moss but in cold weather trucks can be driven across safely."[350] To speed things up further, Paules dispatched Major O. D. Bridges ahead north to the White River with two platoons of soldiers from the 18th, equipped with no less than twelve bulldozers. He ordered them to cross into the fifty-five miles of permafrost and plow north as fast as possible, knocking trees out of the way, and get a bulldozer in contact with the 97th.[351] On October 25 Pvt. Alfred Jalufka piloted a D8 north through the woods far ahead of the two platoons spread out along the trail behind him. Near Beaver Creek, thirty-two miles south of the border, Pvt. Refines Sims piloted a dozer just like it for Company A of the 97th. The two men and their bulldozers met.[352]

The press, in the person of Harold W. Richardson of the *Engineering News-Record*, had come to the border in the nick of time. In his letter of

120   A DIFFERENT RACE

October 26, Parsons told Abbie, "A Mr. Richardson editor . . . for Engineering News-Record was by here yesterday. I took him up the road to meet the 18th (made contact yesterday [referring to the two dozers]). He came back and stayed in my camp all night and I took him to HQ in the a.m."[353] To Parsons and the soldiers on the ground, the meeting of the dozers didn't mean all that much. Notice that in Parsons's letter to Abbie, he merely says in parentheses "made contact yesterday." They hadn't completed the Alaska Highway. The dozers had simply found a way to meet in the woods. Major Bridges would turn his soldiers and bulldozers around to retrace their path back to the White River. The soldiers of the 97th and the civilians who followed them would complete the winter road south to the White River. There they would meet the soldiers of the 18th and complete the Alaska Highway.[354] Two weeks later, in the dry bureaucratese of a progress report to Northwest Service Command, the 97th recorded the actual completion of the highway: "Road constructed on line broken by 18th Engineers from border plus thirty-two miles to border plus fifty-five miles . . . crossing of the White River."[355]

Colonel Paules, General O'Connor in Whitehorse, their bosses in Washington, and, most important, the press, in the person of Harold Richardson, hadn't waited. After Parsons "took him up the road to meet the 18th" and then "dropped him at [97th] Headquarters the next morning,"[356] Richardson posed Jalufka and Sims and their dozers, had them reach across to shake hands, and took the photo that endures as an icon in Alaska Highway history (see Photo 14). And then he filed the story his editor and the army brass wanted:

> In a murky arctic snowstorm at Beaver Creek . . . in the wilds of the Yukon, the climax of building the Alcan Highway was reached at four p.m. Sunday Oct. 25, when the advance tractor crews from east and west came together, closing the last gap in the trail route.
>   After seeing trees fall away from him in weeks of swamping out the advance cut, Corporal Refines Sims, Jr., negro, "cat skinner" of the crew working down from Alaska, frantically, retreated with his diesel bulldozer as trees ahead started falling

his way, not realizing the meeting was imminent. In a few moments, the lead bulldozer of the Yukon crew burst through the last patch of timber and brush, piloted by Private Alfred Jalufka.[357]

The army's publicity machine launched. Secretary of War Henry Stimson didn't wait for Richardson's story. He issued this ludicrous statement to the rest of the press on October 29:

> Trucks started to roll the entire 1,671-mile length of the Alcan highway this week, carrying munitions and material to troops in Alaska. . . . The formal opening will probably take place Sunday afternoon, November 15, at the Alaskan-Canadian border.
> Ten thousand soldiers . . . and 2000 civilian workmen . . . completed the job in slightly over six months. They pushed forward at the rate of eight miles a day, bridged 200 streams, laid a roadway 24 feet between ditches, [and] at the highest point, between Fort Nelson and Watson Lake, reached an altitude of 4212 feet.
> Thousands of trucks will run all winter carrying soldiers and supplies to Alaskan posts. Plans are under way to haul strategic raw materials southward on the return trips.[358]

Stimson had announced a formal opening ceremony, and staff officers in Whitehorse scurried to make one happen. They struggled, first, to choose a location. Clearly, they couldn't transport dignitaries over frozen permafrost to Stimson's location at the border. They chose "the highest point on the roadway built by the 18th Engineers along the western shore of Kluane Lake that summer, the site commanded a breathtaking view."[359] They named the spot "Soldiers Summit" and commenced to make plans. They aimed for Stimson's date, November 15, but climate and geography, as it had so many times before, fought back: "Starting on November 10, a warm chinook wind swept over the southern sector of the highway for four straight days, pushing the temperature as high as fifty-five degrees and causing the one-foot-thick ice on suddenly flooding rivers and streams to break up."[360] As bridge after bridge disappeared in the torrent, General O'Connor postponed the event until November 20.[361]

In Yukon, of course, a significant piece of the road still didn't exist. The soldiers of the 18th still labored to complete the road to the White River, and the soldiers of the 97th still labored south to meet them there. The staff officers in Whitehorse worked at desks next to roaring stoves. The soldiers didn't. Speaking for the men of the 18th, Fred Rust remembered:

> The increasing bitterness of the weather was effecting us more every hour. . . . Swift-moving Yukon streams resisted freezing and the undersides of trucks that crossed them soon became ice coated. . . . Ice would lock the wheels of a truck or car with wet brakes that stood still for a few seconds (not minutes, but seconds), and any attempt to move forward would snap an axle. Sometimes the . . . brakes could be smashed free with a sledge, sometimes gallon cans of burning gas or diesel had to be set under them.
>
> Gravel froze in solid masses in the beds of trucks and men were stationed at the end of the haul to beat it out with picks and sledgehammers. . . .
>
> Vehicles frequently sputtered and stopped dead on the road when water froze in the gas lines. The copper lines had to be disconnected and blown out by mouth. . . . A mouthful of sub-zero gasoline is not exactly tasty. . . .
>
> Trucks used to snake logs through the woods when cats were not available emerged without bumpers, fenders, mufflers, or running boards.
>
> Trucks with bent frames and beds and distorted springs moved crabwise up the road. Some trucks broke in half, [and] were left beside the road as derelicts.
>
> Men suffered but held up better than equipment.[362]

The soldiers of two regiments, still in the frigid woods, had barely made it to the White River when the staff officers had their opening ceremony at Soldiers Summit on November 20. Dignitaries from Canada joined dignitaries from the United States in accommodations at the south end of Kluane Lake: "By the standards of the Yukon wilderness in the middle of November, the accommodations at Kluane Lake were deluxe. Stove-heated, generator-lit, and redolent of spruce planks and tar-paper" barracks housed them in comfort.[363] The next morning at Soldiers Summit, however, the dignitaries got cold. Scurrying staff officers couldn't change the weather, but the 18th Engineers band stood by. Flags fluttered from fresh-cut spruce flagpoles. At 9:30 a.m. the convoy of dignitaries arrived. They heard speeches, and "Alaska's Bob Bartlett . . . presented O'Conner [*sic*] with an Alaskan flag from the Fairbanks chapter of the DAR. . . . Two honored guests jointly accepted a pair of gold-engraved scissors. . . . [They] stepped forward to cut the ribbon amidst rising cheers and the strains of 'God Save the King' and the 'Star Spangled Banner.'"[364] The two young soldiers and their Dodge half-ton weapons carrier who in September had traveled from Dawson Creek to Whitehorse led a convoy of trucks past Soldiers Summit. The band serenaded them as they pulled away up the highway: "By the time [they] reached the bleak outwash valley beyond Kluane Lake, General O'Conner [sic] and his guests were sitting down to a feast of 'moose steak a la Donjek.'"[365] The Corps of Engineers had presented President Roosevelt his land route to Alaska, but "very few vehicles drove all the way through to Fairbanks." (See Map 7 in Chapter 13, on page 100.)[366]

While the dignitaries celebrated and ate, the soldiers of the 97th scattered through frigid Alaska to deal with the most serious problem they had yet encountered: "Col Robinson was irate and worried about winter quarters. The 18th Army Engineers had promised to have portable barracks at the river [White River]. When we got there, there were no barracks. They [the 97th Engineers] were still in tents [at a temperature of −40 degrees]. Food would freeze in their mess kits before they could eat it."[367]

CHAPTER SEVENTEEN
A Miserable Winter

The young Black soldiers of the 97th reached the White River and finished construction of the Alaska Highway. They had given it their all. Now, in November, they turned to face one of the coldest winters on record for northern Canada and Alaska. And what had the army provided? Partially finished, uninsulated wood frame barracks. And tents. Most of the civilian contractors had departed. Some of Duesenberg's men had stayed to help, but they escaped in November and returned to a heroes' welcome in Iowa.[368] But because the army thought it had a brand-new land route to Alaska and the army intended to use it, the soldiers of the 97th would stay to keep it clear of snow and ice and keep it in repair through the winter. Northwest Service Command proposed to keep the truck supply convoys, called "the Fairbanks Freight," rolling right through the winter. General O'Connor ordered the 97th to establish way stations, or terminals, through Alaska and to keep the new highway open and passable from Beaver Creek north.[369]

Colonel Robinson dutifully distributed his soldiers along the way from the White River all the way north to Big Delta. But the soldiers on the ground knew a truth: the highway from Whitehorse north to Big Delta wouldn't support the Fairbanks Freight this winter. Snow plugged the road; ice mushroomed across it; crude, hastily built bridges collapsed. At the Robertson River, between Tok and Big Delta, ice formed seven feet above the bridge that the civilian contractors had built; the bridge itself was nineteen feet above the streambed.[370] "Very few vehicles drove all the way through to Fairbanks. Although the road was now officially open from end to end, it was still really only an emergency supply route—particularly northwest of Kluane Lake—and a very tenuous one at that."[371] Every man in the 97th, white officer and Black enlisted, knew another truth: they had to survive here until spring. They struggled through the bitter winter to build housing for themselves. And the weather conditions of winter pinched their supply lines.

On November 1 Parsons wrote to Abbie with an update: "We have had a big change in the setup since we met the 18th. Only two companies are on the old jobs, the balance [are] on a building program, on a belated building program, since it's often below zero and the snow falls every day now. It's far from tenting time."[372] On November 9 he followed up: "Most of the fellows, officers anyway, think the Colonel has let us down by not having someplace built for us to get in out of the cold."[373] On November 18 he apologized to her for a missing page from his letter—he had left it on his desk when he went to dinner, and snow blew through a hole in his tent and soaked it.[374]

Before they departed for the warmth of home in November, civilian contractors had worked on barracks at Big Delta, Big Gerstle, Tanana Bend, Salchaket Lake, and Cathedral Bluffs, which by November 20 were 85 percent complete. The H&S Company wintered at Big Gerstle, and the PRA civilians had uninsulated barracks in place there by November 20.[375] Company C, wintering at Cathedral Bluffs in the PRA-built barracks, struggled to complete them. By December Company B had made their way to the area around Robertson River between Tok and Big Delta. The departed civilians had left construction shacks behind. The shacks had been constructed, however, to house men in summer, not winter.[376] The soldiers of Company D and their sawmill had moved up to Midway Lake. The carpenters of Company E moved to join them. By October 24, Cpt. James Fischer reported five barracks nearing completion there.[377] Company A settled at Beaver Creek in November, and Captain McMeekin directed Lt. Walter Mason to begin building log cabins. Mason described his plans: "I drew

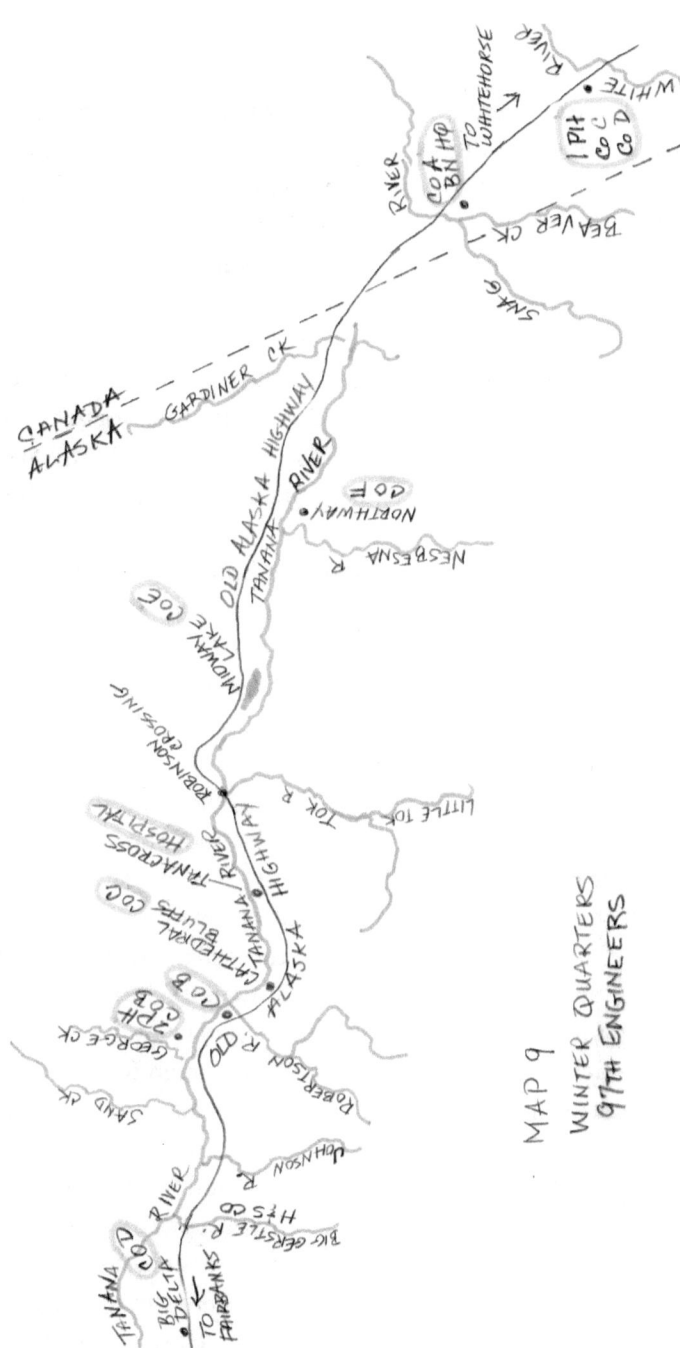

Map 9 Winter Quarters

a proposed design of log cabins, roughly twenty feet square, with walls about seven and a half feet high, with sawed wood floors, roofs, windows and doors."[378] Log cabins, of course, require logs. Company D dispatched a platoon and a saw to help acquire them. Temperatures already hovered around twenty below and lower, and the outer surface of the spruce trees Mason's men proposed to use had frozen to a "depth of two to four inches, making it necessary to slab [saw] off this frozen portion."[379] Mason's men labored through November and December during each day's five or six hours of daylight. When the temperature got colder than forty below, they stopped, "and there were a number of those days."[380] While they labored, they lived in tents. The cooks prepared meals in a sled-mounted cook shed inherited from Duesenberg.[381]

At the settlement of Northway, Parsons's Company F had big plans—seven 20 × 52-foot buildings, a 104 × 20-foot mess hall with a 20 × 20-foot kitchen wing, latrines, and a hospital. But progress came hard. On November 11 they recorded a temperature of zero degrees and reported all the buildings "laid out." Two days later, on November 13, they reported two buildings complete with roofs, three buildings with sills and subfloors, and two with foundation posts. They had also excavated a root cellar. On November 20 they recorded three inches of snow and a temperature of ten degrees below zero. Four buildings were now complete except for tar paper, which served as siding and insulation. Parsons and most of his soldiers remained in tents.[382] Captain Parsons explained the lack of progress to Abbie in his letter of December 7: "I gave my company the day off. . . . There's not much we could do anyway because we are out of stuff to do it with. We have started a camp here big enough for 100 men."[383]

The regiment didn't just struggle to get building material. It struggled to get supplies of any kind. Knowing that Thompson Pass would soon close for the winter, soldiers and civilians had raced to get as much material as possible out from Valdez. In November the pass closed. If something they needed hadn't made it to Slana by then, it wouldn't come out from Valdez until spring. On December 12 Parsons messaged headquarters: "We have only three days rations. . . . How about sending us something to eat." On December 15 he sent an update: "We will be out of rations tomorrow."[384] By the end of December, he was losing patience. On December 30 he messaged, "Regt Supply Officer. . . . We had chili con carne for Christmas, how was yours?"[385] Many years after the fact, Parsons's son remembered his dad quoting one of his messages: "Send food or send coffins."[386] Not surprisingly, some of the soldiers got sick. On December 28 Parsons messaged Colonel Robinson: "Have sent several messages by truck regarding sick EM [enlisted men]. . . . We have several very sick men and no aid men or medical equipment at this camp. We should have a medical officer come pay us a visit."[387]

Infinitely worse than keeping the young Black soldiers in Alaska without housing, the army kept them there with totally inadequate clothing. Their uniforms, ineffective in subarctic conditions to begin with, had worn out. On December 9, 1942, the army asked H. Bradford Washburn to travel to Alaska and inspect the uniforms and equipment provided to the soldiers of the 97th. Washburn, a mountain climber and expert on cold weather gear, had worked with the Quartermaster General Corps to conduct cold-weather clothing experiments.[388] He traveled to Alaska, and what he found in January 1943 horrified him:

> A thorough survey was made of clothing and equipment of the 97th Engineers between Fairbanks and Northway. Temps to sixty-three degrees below zero were encountered in the field, and clothing of this unit was found to be in abominable condition, so much so that specimens of it were brought to Washington, DC, to illustrate the extremes under which American troops are operating in the field when their supply has been neglected.[389]

Washburn summed it up this way: "The pathetically ill-equipped 97th Engineering Regiment on the northern quarter of the Alcan Highway is doing little else but hibernating at present."[390]

Mary Hansen confirmed Washburn's conclusion. A hundred young Black soldiers from the 97th wintered near the roadhouse that she ran with her husband, Bert, at Big Delta. An old double cabin served the unit as a field kitchen, but the men lived in frigid tents. At Christmas, Bert made up a package—a comb, cigarettes, a candy bar, soap—for each man. When Mary went with him to deliver the packages, the tattered condition of their uniforms horrified her as much as it had horrified the army's clothing expert, Washburn. It was bad enough that the men were living in tents through an Alaska winter, but it was infinitely worse that they worked outside in uniforms so ragged that she could see their skin.[391] Tech 5 Allen G. McCray told his daughter about the clothing: "There was not enough, and it wasn't thick enough. We shared clothing with each other."[392] However, the regiment saw to it that white civilians fared significantly better. On November 7 Duesenberg's men, a couple of weeks from going home to Iowa, shivered behind the 97th on the way to the White River. They requisitioned and received from the regiment forty "Parka[s], Alpaca Lined."[393]

Cpt. Richard L. Neuberger, an aide to General O'Connor, penned an article for the *Saturday Evening Post* in 1944 that vividly portrayed the cold of that awful winter. "Crack an egg," he said, "and you found it full of ice crystals. Potatoes developed ribbed strips that looked like Italian marble."[394] Jean Louis Coudert had served as Catholic bishop of the Yukon for many years, and in 1942 he visited army camps up and down the Alaska Highway. He told Captain Neuberger that he "watched a hot dinner freeze on his plate as he rushed the 120 feet from the kitchen to the mess tent. . . . Chunks of ice were floating in my shaving lotion one morning when the mercury touched 61 degrees below."[395] More than anything, the soldiers appreciated warmth whenever and wherever they found it. Neuberger explained:

> Outdoors at almost 100 degrees below the freezing point [sixty-eight below zero], your face feels as if it were being branded with a hot iron. Knees encased in woolen underwear quaver involuntarily. Feet tingle painfully and then become numb. One yearns for warmth. . . . I remember the colored staff sergeant who grinned at me as we sat in a pyramidal tent on opposite sides of a glowing little wood stove that had been fashioned from a discarded oil drum. . . . "Ah's happy, lieutenant," the sergeant said, "'cause ah's warm."[396]

Other soldiers of the 97th, interviewed years after the fact, had especially vivid memories of the cold. Clifton Monk of Company B remembered, "Lord it was bitter cold. Your breath would turn to ice inside your blankets at night. If you touched anything with your bare hands, you couldn't tear your skin loose."[397] William Griggs, the regimental photographer, remembered this for a PBS interviewer: "It was actually a fight for survival. In one case, I spilled some gasoline on my clothing, and it evaporated. And I went to pull my coat off, it pulled the flesh off too."[398] Jesse Balthazar explained to John Virtue, "The boys found it better just to wear plain galoshes and wear two or three pair of socks. That's what just about everybody would wear. . . . You can tell how cold the temperature is by the number of eyelets that can be laced before the fingers can no longer hold the shoelaces. After you warm your hands under your armpits, you can continue the lacing of the boots."[399] The daughter of Tech 5 McCray, who served with Company D, remembers stories he told about the winter: "Their breath froze in front of their face." He also told her of icicles hanging off his hair, eyelashes, and eyebrows.[400]

Fighting an implacable enemy, the regiment took casualties. In January 1943 Lt. Allen Lytz of Company E set up a platoon camp at Clearwater Creek. A few days later he sent Tech 5's Joseph S. Smith and K. V. Nelson out on a routine mission. As they drove their truck past Station Creek, about thirty miles from Clearwater Creek, it slid into a ditch, broke through a layer of ice, and stuck fast. Thirty miles from salvation, the two men had no choice—around 11:00 a.m. they started to walk. Ten miles on they stopped and built a fire so that Nelson could warm his feet, but they had to keep moving. After a few more miles, they stopped again. This time they couldn't build a fire. They had matches, but their freezing hands couldn't grasp to light one. As they trudged on through the night, cold and fatigue built up for Nelson. He fell several times. Smith managed to get him back on his feet, but it got harder and harder to do. Smith had to leave him, hoping against hope that he could get to camp, find help, and get back in time to save the soldier. At 2:00 a.m. Smith stumbled into camp, and four soldiers set out to find Nelson. At 8:00 a.m. they found his frozen corpse.[401]

On December 18, 1942, Captain Parsons wrote to Abbie that the officers in Company F had abandoned their tent and moved into the root cellar, "a log structure 16' × 24' which is down in an 8-foot hole with 2 feet of dirt and rock over it. Until today all I had down here was my bedroll and bed. I want to get the balance down today but its just –48 degrees [48 below zero] outside and you don't do much moving when it's that cold." His troops, of course, stayed in their tents.[402]

For the young Black soldiers of the 97th, the endless, agonizing winter passed very slowly. Hibernating in their tents, they survived by looking out for each other, by having each other's backs. They shared clothing and watched each other for telltale signs of frostbite. And they endured.

Photo 1 Valdez existed as the port of entry for a transportation system that served the rugged northern interior of Alaska. In other words, Valdez connected the oceans of the world to the Richardson Highway. *Source: National Archives.*

PHOTO 2 The USS *David Branch* sailed into the Valdez Port arriving at the narrow wooden dock and the warehouse. The soldiers disembarked with duffle bags and marched along the Alaska Avenue passed framed buildings of Valdez to intersect the Richardson Highway. *Source: National Archives.*

PHOTO 3 Along the west bank of the Slana River lay the Slana sand hill. The catskinners started the road along this hill and their D8 dozers tended to slip sideways instead of bulldozing forward. *Source: National Archives.*

PHOTO 4 At about 5:50 am, the eye of the storm passed over Dutch Harbor. The rain and fog cleared just as the Japanese planes descended into the attack. They bombed the base and the harbor for the next twenty minutes. *Source: National Archives.*

PHOTO 5 A dozer pulls a log sled and two carts loaded with the soldiers personal belongings to their next bivouac. *Source: National Archives.*

Photo 6 A catskinner's dozer pulling a carryall is stuck in the mud. Soldiers carried logs using them as corduroy to give a second dozer more traction to pull the first dozer out of the mud. *Source: National Archives.*

PHOTO 7 Soldiers with hand tools scrambled over the right of way cutting and clearing trees. *Source: National Archives.*

PHOTO 8 The solution to endless deep mud and muck was corduroy. The big cats slewed through once, knocking trees down, then the soldiers on the ground would cut thirty-foot logs and drag them to lay side by side across the right of way. *Source: National Archives.*

Photo 9 The soldiers build culverts, a square tunnel, from rough logs and timbers for the stream to flow through. *Source: National Archives.*

PHOTO 10 They did not have concrete or steel, of course, so they built bridges of timber, driving pilings every twenty feet or so and spanning them with timber stringers. Across the stringers they laid planks, decking, for the roadbed. *Source: National Archives.*

PHOTO 11 At the Tanana River, Company A had a problem. Pontoon boats and bridging material had not arrived yet. Company commander McMeekin hailed a sternwheel steamboat and quickly made a deal with its owner. *Source: National Archives.*

Photo 12 The soldiers managed to wedge a little frivolity into the schedule. Back in August Captain Parsons bought his son a Husky puppy, named him Tok. *Source: Parsons Private Collection.*

PHOTO 13   The G.I. stove never reached this unit so they used open fires and an oil barrel oven. *Source: National Archives.*

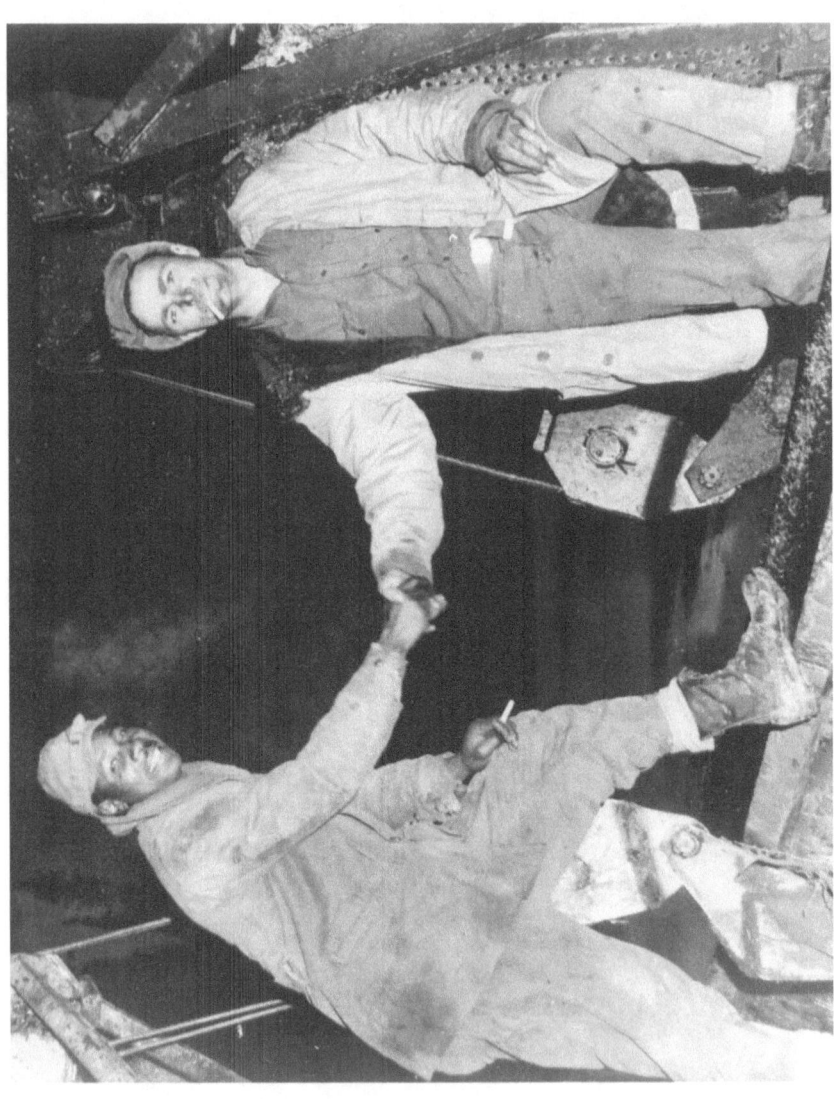

PHOTO 14 Harold Richardson of the *Engineering News Record* posed Sims, 97th Engineers, and Jalufka, 18th Engineers, on their dozers, had them reach across to shake hands, and took the photo that endures as an icon in Alaska Highway history. *Source: National Archives.*

CHAPTER EIGHTEEN

Black Troops Need Discipline

The 97th had struggled through the spring and the first half of the summer of 1942 and had all but fallen apart in disorganized confusion. In August and September Colonel Robinson had reinvigorated the regiment, reorganized it, then reorganized it again. And the 97th had beaten the 18th to the border. In November they completed the winter road all the way to the White River and began to reposition and prepare for a new mission. General O'Connor and his officers at Northwest Service Command, believing they had a highway all the way to Fairbanks, planned to move freight over it through the winter. They ordered Robinson and the 97th to keep it open and passable.[403]

The National Guard colonel had found a way to galvanize his regiment and complete the road. But his regiment had serious organizational problems. The rush to the White River had temporarily buried the problems; but now, as Robinson's officers turned to face the winter and a new mission, the problems refused to stay buried. Three issues came to the surface, one after the other, through the winter, and by the end of January the regiment had dissolved once again into disorganized and demoralized confusion. The turmoil, for the most part, occurred among Robinson's officers, but it profoundly affected the young Black soldiers who worked for them as well. First the officers of the 97th thought about their careers and their prospects for promotion. Commanders and senior officers at the top of the food chain, the men who held the keys to promotion, still didn't think much of or expect much from the segregated 97th. They ignored it if they could. An officer looking to display his talents and attract favorable attention from those officers needed to find a better place from which to do it. To be sure, the 97th had beaten the 18th to the international border and had built the highway into Canada all the way to the White River. But General O'Connor and his staff at Northwest Service Command saw that as a fluke. "This outfit," Parsons explained to Abbie in his letter of November 1, "was on the wrong end of the road for supplies and now when the General [Hoge] pulled out was on the wrong end for good jobs. My guess is it will always be on the wrong end."[404]

So in November many of the regiment's best officers maneuvered to get out of the 97th and into more promising jobs elsewhere. A National Guard officer, at heart a civilian, Robinson didn't understand or know how to react to the officers' desire to get away from his regiment. In a November 1 letter, Parsons explained to Abbie what happened when he had asked the colonel to help him get a transfer:

> It made him sore at first, but after a bit he said he would help me. I don't think I rate very high with Robinson but he is kinda on the spot and is not too sure of himself and the idea of officers wanting to get away from the 97th makes him wonder. I told him my reason. . . . Gordon and Schneider also want transfers—Goglia has gone. . . . You can see why the Colonel expects something. He is almost right . . . in what he expects because most of the officers in all the other regiments up this way are getting promoted while our gang sits pat. Everyone places the blame on the Col. because he hasn't even been able to get himself promoted while all the other regimental commanders are getting themselves good jobs. The 18th Colonel took over Hoge's job. The 18th Lt. Col. took over the 93rd and my friend Bridges who was a Captain not long ago may become Lt. Col. soon.[405]

Parsons spotted his own opportunity for escape in something called the Canol Project.

In early 1942, planning the Alaska Highway project, generals in Washington had realized that truck convoys over the new road would not accomplish much if they had to use most of their hauling capacity to haul their own fuel. They decided to install a six-hundred-mile pipeline from the Canadian Oil (CANOL) fields at Norman Wells to a new refinery in Whitehorse. As the Corps of Engineers rushed, through 1942, to complete the highway, the Canol Project had sat on a back burner. But in early 1943 General Somervell made it a top priority.[406] The Office of the District Engineer in Whitehorse would need experienced officers, officers like Cpt. Walter Parsons. In his November 1 letter to Abbie, Parsons told her about a visit from "the Padre," a Catholic priest based in Whitehorse: "I had a talk with him about the job I'm trying for and he's going to see what he can do about it. . . . He is not so hot for Robinson (nor am I) and I understand he got Goglia his job in Whitehorse."[407] It would take Parsons awhile to make the right connections and seize the opportunity in Whitehorse, so he endured the worst of the winter with Company F.

The second issue for the officers of the 97th came to the surface in November and early December as Robinson moved his companies into position along the icy, snow-covered road through Alaska. When the companies arrived at their winter locations, their officers got their first look at the miserable accommodations intended to house and feed them and their men through an Alaska winter. Many of them would struggle to build their own rough barracks and many of them would winter in tents.[408] The angry officers, including Captain Parsons, blamed Colonel Robinson. On November 9, Parsons wrote to Abbie, "Most of the fellows—officers anyway—think the Colonel has let us down by not having someplace built for us to get in out of the cold. He was told plenty of times when it would get cold and how cold we could expect it to get. It looks now like we might have to build log cabins for ourselves."[409]

The third issue involved the mission itself. Northwest Service Command had left the men of the 97th in Alaska to keep the highway open so that truck convoys to Fairbanks could use it. However, they had positioned other regiments along the highway through British Columbia and Yukon with the same mission. Against daunting odds, the other regiments succeeded and truck convoys made it as far as Whitehorse, but very few vehicles moved on from Whitehorse through Alaska to Fairbanks. The Tanana Valley in Alaska served up especially bitter and sustained cold: "The Upper

Tanana Valley . . . gets very cold; it is about 1000 feet higher elevation than the Yukon Flats."[410] The valley occasionally suffers strong winds as high as eighty miles per hour, and the winds take the wind chill eighty or even one hundred degrees below zero.[411] Winter weather in the valley took the job of keeping the road open from daunting to impossible. Clearing snow from the road presented a problem, but in Alaska mounding ice presented a worse problem. The road through the Tanana Valley crossed rivers and stream after stream, and in late summer and fall the regiment and the civilians behind them had installed culverts so that water could flow under the road, or they had bridged rivers and streams with rough timbers. The bitter cold of winter ruined everything.[412] Even the arctic cold couldn't form ice in a swift-moving current, and the water continued to flow. But ice formed along the sides and the bottom of the flow, gradually compressing its channel, raising the water level until water flowed over a culvert, over and around a bridge. Great mounds of ice formed and grew to block passage. Because the continuously growing and expanding mounds of ice seemingly mushroomed out of nothing, the soldiers called these mounds "mushroom ice."[413] Snow and mushroom ice blocked the road at every turn. And "temperatures dropped to 72 degrees below zero on the upper end of the road. . . . In the latter part of January all outside work ceased."[414] Hibernating soldiers couldn't do much about snow or mushroom ice.

Gradually the officers of the 97th realized that Northwest Service Command had assigned them an impossible mission. They and their men were enduring the misery of an Alaska winter for no reason. Morale sank to an epic low. The frustration and despair of the regiment didn't escape the attention of General O'Connor and his officers at Northwest Service Command. If they had tolerated Colonel Robinson's unorthodox organization in 1942 because it got results, they had no reason at all to tolerate them now. When, in January, word of Washburn's devastating verdict on the clothing of the soldiers of the 97th reached Whitehorse, they didn't have to look far for someone to blame.

Captain Parsons finally got his transfer to the Office of the District Engineer and left Company F to report to Whitehorse on January 16, just as the situation for the 97th came to a head.[415] On January 24 he wrote to Abbie that some "special service men" had been up with the 97th assessing morale. And they had brought back a "report that morale is low, and some big shot wants to know why. . . . I understand Mitchim is in town, but I have not seen him."[416] The general would normally have summoned the regimental commander, Robinson. On February 3, 1943, the general made clear why he had summoned Colonel Mitchim instead. He removed Colonel Robinson and placed Colonel Mitchim in command of the 97th.[417] For the officers and soldiers of the regiment, everything changed once again.

Mitchim had served with the 97th since Camp Blanding, which meant he had served through all the turmoil from Valdez to the White River. Like General O'Connor, Mitchim had gone along with Robinson's radical reorganization in September because it worked, but as a West Pointer and a career army officer, Mitchim had worried about the impact of the reorganization on the regiment.[418] The events of the winter had confirmed his misgivings. In February Mitchim had multiple issues to address, but doing so would require an efficient, motivated organization. The army had taught the West Pointer that efficiency, morale, and motivation depended, first and foremost, on military organization and discipline. The army had also taught a set of assumptions about Black soldiers, a set of assumptions that rendered discipline in a segregated regiment especially important—and especially difficult to maintain.

In his *The Employment of Negro Troops*, Ulysses Lee quotes a study produced by officers at the US Army War College that described the "personality problems which commanders expected to meet":[419]

> As an individual the negro is docile, tractable, lighthearted, carefree and good natured. If unjustly treated he is likely to become surly and stubborn, though this is usually a temporary phase. He is careless, shiftless, irresponsible and secretive. He resents censure and is best handled with praise and by ridicule. He is unmoral, untruthful and his sense of right doing is relatively inferior. Crimes and convictions involving moral turpitude are nearly five to one as compared to convictions of whites on similar charges.[420]

A War Department memo on the use of "American Negro troops" in Canada, and, by extension, in Alaska, concluded, "Negro troops should be subjected to rigid discipline. Negro troops should be officered by white men."[421]

Among Mitchim's officers, 1st Lt. DeWitt C. Howell had graduated from the Carnegie Institute of Technology in 1940 and accepted a commission as a second lieutenant in the Army Reserve.[422] Summoned to active duty, he had reported to the 97th in Florida and traveled to Alaska with Company E. He assumed command of the company in May. Promoted to first lieutenant in July, he had remained in command of Company E through the rest of 1942.[423] Lieutenant Howell had plenty of experience as a company commander, but, crucially, he hadn't had the same experience in the field as Captain Parsons. When Colonel Robinson reorganized three companies of his 1st Battalion, the officers, like Parsons, who commanded those companies had adjusted to a new reality. Working especially closely with their Black soldiers as they turned them into civilians in uniform, they had got past stereotypes and had begun to see them not as Black men but simply as men. These officers agreed with their fellows on the importance of military organization and discipline, but they didn't see extra difficulty or a greater threat from Black soldiers. But the reorganization hadn't affected Lieutenant Howell's company. Company E, handling supplies and building buildings at Tok, had continued to operate as a normal company. Lieutenant Howell had still accepted the army stereotypes about his Black soldiers, and he continued to expect *all* of his soldiers to stand formation, salute, and respond to orders without question.[424]

Second Lieutenant Robert W. Lyon, from New Britain, Connecticut, had graduated from high school and moved on to art school in New York when the army came calling. Officer Candidate School (OCS) had turned young Robert into an officer.[425] Barely out of his teens, with no military experience, he arrived at the 97th in January 1943. The regiment did what regiments do with brand-new second lieutenants, referred to as "shavetails": they put him in the H&S Company, where, sooner or later, someone would find something for him to do.[426] On March 3 Captain Sprigg, commander of the H&S Company, transferred out of the regiment, and Colonel

Mitchim sent Lieutenant Howell from Company E to replace him.⁴²⁷ At his new command, Lieutenant Howell found the inexperienced young Lyon and created a job for him that would keep him close. He called Lyon his "Assistant Commanding Officer."⁴²⁸

In the spring of 1943, Noah Williams, a Black sergeant from Lenoir, North Carolina, had served Captain Sprigg as his first sergeant in the H&S Company for eight months. But Sergeant Williams lucked into a leave in early February and returned on March 25 to a new H&S commander, Lieutenant Howell.⁴²⁹ A company commander works closely with his experienced first sergeant, or "top kick." They form a unique relationship. The first sergeant provides the commander with eyes and ears and exercises a significant amount of the commander's authority. A top kick would normally keep an eye on a callow young shavetail; he would guide him and would let the commander know if the shavetail took a wrong turn. Lieutenant Howell and 1st Sergeant Williams would eventually form a relationship, but they didn't have one on March 25. Sergeant Williams would not be helping Howell keep an eye on young Lieutenant Lyon.

In Washington, Gen. Brehon Somervell, the man at the very top of the chain of command for the Corps of Engineers, had a notion that the army needed to support the lend-lease transfer of warplanes from the United States to Russia by building another highway from Fairbanks up to the airfield at Nome, Alaska.⁴³⁰ Orders trickled down the chain, and in March men from the 97th began moving north of Fairbanks to settle at Livengood, Alaska, preparing to build Somervell's road.⁴³¹ The regiment needed a supply depot at Fairbanks. When Lieutenant Howell's H&S Company received routine orders to establish the depot, Howell spotted a job he thought Lieutenant Lyon could handle and ordered him to plan and manage the effort. The supply depot would require soldiers, of course, so Lieutenant Howell turned to Company F with a request to supply some men. Sergeant Heard's boss, the first sergeant of Company F, had used the word "detached" when he told Sergeant Heard that he and his squad would be moving to Big Gerstle to duty with the H&S Company. That would be temporary, they would remain part of Company F.⁴³²

On March 10, Sgt. James M. Heard, Pfc. Lee I. Ratliff, Pfc. Eugene Fulks, Pvt. Sims Bridges, Pvt. Willie L. Howell, Pvt. Robert M. Rucker, Pvt. Warren H. Lindsey, Pvt. James V. Hollingsworth, Pvt. Willie B. Calhoun, and Pvt. Josh Weaver woke up in their tent barracks at Northway Junction. They donned their worn and ragged uniforms, packed their barracks bags, tossed them into the back of an army truck, and climbed in after them, arranging themselves as comfortably as they could among the bags. The weather, thankfully, had warmed a bit since the dreadful days of deep winter, and the temperature hovered just above zero.[433] As the truck rumbled away from Northway, though, the wind of its passing created a wind chill effect. The ten men in the bed of the truck, clad in uniforms never intended for extreme cold, uniforms that time and rough work had worn out, endured an effective temperature well below zero. And the miserable, 138-mile trip to Big Gerstle lasted two days.

On March 12 the truck rumbled slowly into a scatter of rough frame buildings in the woods along the Big Gerstle River and stopped next to H&S Company headquarters.[434] The ten soldiers slung their barracks bags down and clambered out of the truck bed. They walked about, stretching out kinks and stamping their feet, trying to warm themselves. Sergeant Heard reported to the company commander, Lieutenant Howell. And Howell turned them over to the very young 2nd Lieutenant Lyon. Lyon had them pick up their barracks bags and directed them to their new quarters. At least they would sleep in a frame building instead of a tent. Beyond that they had little idea what the young lieutenant had in mind for them.

The men had come a long way from the "invisible hand" South where they had all grown up.[435] In the army they had glimpsed possibilities for themselves that hadn't occurred to them before. And the army trained them, gave them skills. Crucially, the army had picked Sergeant Heard for NCO school back in Florida, where he had trained to lead and supervise his men by looking out for them and taking care of them. Experience on the road and the bitter experience of their winter in tents at Northway had prepared them further and honed their skills. When the hurry to finish the road and Robinson's reorganization turned them into civilians in uniform and formations and saluting stopped being important, in August, they barely noticed. They worked hard, very hard. They got good at their jobs. Then came the bitter experience of their winter in tents at Northway. If Sergeant Heard looked out for the squad and had their backs, the squad had also learned to look out for each other and have each other's backs.

Gradually the weather and the cold had eased, and their misery eased along with it. But through the winter they had gained one more skill—survival. They came out of the winter knowing two things: first, they could take damned near anything the army threw at them and find a way to survive it; second, the army and some of its inexperienced young white officers would continue to screw up and put them in jeopardy. If surviving meant resisting stupid orders, they would find a way to do it. In their frame barracks at Big Gerstle, they waited cautiously for what would come. They had no idea that Lieutenant Lyon and, more important, the company commander he reported to, thought that as Black soldiers they presented a special discipline problem. The stage was well and truly set for conflict.

CHAPTER NINETEEN
Events at Big Gerstle River

On the morning of March 29, 1943, at Big Gerstle, the sun rose slowly behind the trees, gradually illuminating the crude frame buildings of the army camp scattered through the woods. Specialist Willie Stallworth's bugle sounded clear in the frigid air, and in the barracks Sgt. James Heard and his squad slowly came awake. A few minutes later they reluctantly emerged from the barracks into air that had returned to vicious cold. Big Gerstle saw a low temperature that day of 34 below zero.[436]

Lieutenant Lyon had worked hard on the details of the move to Fairbanks, and for several days the Black soldiers had worked for him. They had accumulated and stacked supplies near the H&S Company headquarters: drums of gasoline and diesel fuel, stacks of firewood, boxes of rations, tents, tent stoves, and stovepipe.[437] Neither Lyon nor anyone else had found it necessary to share the details of the planned move with Sergeant Heard and his men. In Colonel Mitchim's regiment, officers planned while Black enlisted men followed instructions. But Heard and his men knew the ropes; they understood that the young lieutenant planned to move them and the piles of supplies somewhere.

The previous autumn's rush to the White River followed by the long winter had left the regiment's rolling stock in sorry shape, but Lyon had managed to scrounge four worn-out Studebaker trucks, and on the morning of March 29, as he sat at his desk in headquarters, he heard the whistling rumble of engines as the trucks rolled up and parked next to the stacked supplies outside. Emerging from the relative warmth of the barracks into the frigid Alaska morning, Heard and his men nervously eyeballed the raggedy old trucks. Only one had a canvas cover, worn and torn. Ice and packed snow mounded in all four truck beds. Exactly how far did the young lieutenant plan to haul them?

Just a few days earlier, Sergeant Heard and his men had traveled in the back of a truck from Northway to Big Gerstle during an Alaska "warm spell." Warm spell or not, the wind chill effect in the moving truck had made the truck bed dangerous and thoroughly miserable. Now the warm spell had ended. They could only imagine the wind chill effect today. If young Lieutenant Lyon didn't know the danger, the other men sure did. And young lieutenants didn't have to ride in the back of trucks.

The squad reported to Lieutenant Lyon in the company area and went to work. Scraping the ice and snow out of the way as best they could, they loaded supplies into the three uncovered trucks. The bitter cold numbed their fingers and toes, and the men built a fire in the yard, periodically retreating to it to warm themselves. Standing next to the fire, they glanced repeatedly at the fourth truck and its worn and ragged canvas cover, worrying aloud among themselves. With the other three trucks loaded and ready to go, the men broke for lunch and took the opportunity to warm up—and confer—back in the barracks. Sergeant Heard had finally learned their destination and he shared it with the others: the damned lieutenant intended to haul them 130 miles to Fairbanks!

Pvt. Willie Calhoun headed across to the regiment's personnel office to work out the details of an allotment from his paycheck. The rest of the men remained with Heard. They had to do something to stop the trip. Traveling so far in the severe cold could hurt them, even kill them. The army had trained Sergeant Heard to his responsibility for his men, and when Lieutenant Lyon, ready to load up and go, made his way to the barracks, Heard hesitantly, carefully, asked him about the truck. He pointed out its bedraggled condition and the danger of riding in it to Fairbanks. Unsure of himself, Lyon left the men for a few minutes and walked back to the orderly room to consult with Lieutenant Howell. Soon he emerged from the orderly room, reassured and back in charge, and ordered the men out of the barracks. The truck would do just fine, he said, and Heard and his men needed to stop stalling, load their personal gear, and climb aboard.

Heard and his soldiers had understood since childhood that Black men didn't argue with white men. The army had trained them that enlisted Black soldiers didn't argue with white officers, even white officers barely out of knee pants. But disaster loomed in the back of that deuce-and-a-half. Experience had trained them not to just lie down in the path of disaster.

Sergeant Heard formally requested Lieutenant Lyon's permission to speak with Lieutenant Howell. Still confident, Lyon granted permission and then followed the sergeant into the orderly room, where Heard tried to reason with the company commander. When he pointed out the condition of their uniforms, Lieutenant Howell responded that "he was not a boy scout but in the army." Heard went on to point out that riding in a truck that was in such poor condition threatened his men with serious injury or death and that he wouldn't stand by and see that happen, wouldn't be responsible. Incredulous, Howell listened to the Black enlisted man argue with him and saw exactly what he and his fellow white officers had long feared they would see. The Black sergeant was questioning his authority, and Howell could not allow that. He informed the recalcitrant sergeant that responsibility fell to white officers, not to Black NCOs; then he curtly ordered Heard to go out and get his men on the truck.

First Sergeant Noah Williams had gone to check on a crew sawing firewood; when he returned, Lieutenant Lyon informed him that they had a "mutiny in the ranks." Not surprisingly, neither his commander, Lieutenant Howell, nor young Lieutenant Lyon had consulted with 1st Sergeant Williams about the planned trip. Heard had rejoined his scared and angry men, milling about and complaining outside. Williams walked over to talk to them. Horrified but out of the decision-making loop, he couldn't help but offer his fellow Black soldiers an experienced NCO's opinion. The men would be "damned fools" to get on that truck. Further, he informed them, standing orders from Whitehorse expressly forbade riding in the back of an unheated truck at thirty below. Caught in the middle between the Black soldiers and the white commanders, and probably realizing that by offering his opinion he trod on thin ice, Williams moved to escape to his desk in the orderly room.

At about the same time, hearing voices, a commotion, outside his orderly room, Howell reacted, determined to get the situation under control. He stalked to the door, yanked it open, and ordered the men to get onto the truck. When they didn't immediately comply, he upped the ante by ordering them to line up and come, one at a time, into the orderly room. As each man entered, he made a point of recording his name, rank, and serial number; ordered Lyon to make a duplicate list; and asked each man whether he intended to follow orders and board the truck. For at least one soldier, the danger and the stupidity of the order trumped his fear of the white officer. Pfc. Sims Bridges flatly refused to get on the truck. When Howell countered that Bridges had already been in trouble and that he, of all people, should want to comply, Bridges responded that he didn't care. He didn't propose to freeze to death.

Sergeant Heard and most of the men chose to stick together, answering that they would board the truck if the others did. The scared and angry soldiers huddled outside, arguing about what to do, and Howell sent First Sergeant Williams to put them in formation. Young Lieutenant Lyon went out to talk to them as well. Resisting an officer's orders amounted to mutiny, he explained. And the army punished mutineers by executing them. He strongly advised them to climb onto the truck.

Lieutenant Howell and his fellow white officers had expected discipline problems from the Black men in their charge; they had known something like this incident would happen sooner or later. Now, cursing the luck that had laid it at his door, knowing his career was at stake if he handled it wrong, Lieutenant Howell emerged from the office to play his trump card. Walking to a spot between the assembled soldiers and the back of the truck, he delivered a short speech. He would issue a clear order. A man refusing it would be committing mutiny. The army put mutineers to death. He raised his arm, fixed his gaze on his watch, and gave them ten seconds to fall in behind him if they intended to go. Four of the men moved, not exactly behind him but clearly intending to comply. With his eyes fixed on his watch, Howell didn't notice. Ten seconds elapsed and Howell stalked away to regimental headquarters to confer with Cpt. Jack Doyle, standing in Colonel Mitchim's absence as temporary regimental commander. Howell proposed to cancel the trip and put the men under arrest in quarters. Doyle agreed. By the time Howell returned to the H&S Company area, all ten soldiers had boarded the truck, but Howell ignored that they had eventually complied. He brusquely ordered Lieutenant Lyon to get them off the truck and return them to the barracks—arrested in quarters.

Fate had chosen Howell to deal with this situation, and he knew his career hung at risk. He knew his backside needed covering. When the regiment moved to make an example of these men, everything he had done would come under scrutiny. First he needed to be sure he had the names right. He didn't really know these men at all. Both he and Lieutenant Lyon had listed names and serial numbers when the men came into the orderly room, but now he couldn't find his list, only the one made by Lyon. He dispatched his company clerk, Private Hinson, to the barracks to make another list. Hinson's list included Private Calhoun, who had finished at the personnel office and returned to join the squad. Mysteriously, though, Lieutenant Lyon's list included Calhoun too, even though he hadn't actually been there.

Howell also hadn't paid any attention to what the men wore that day; he had dismissed Heard's concerns about uniforms. But every officer in the 97th knew that Washburn's report had made the condition of the soldiers' winter clothing a hot-button issue for the regiment. Howell dispatched Lieutenant Lyon to inventory the ten soldiers' clothing and then joined him in the supply room to hurry the process. Not surprisingly, according to the lists prepared by the two young officers, all ten had plenty of adequate clothing. Their uniforms wouldn't have presented a problem for the trip to Fairbanks. Strangely enough, when the ten arrived at the Whitehorse stockade a few days later, wearing and carrying exactly the clothing Howell and Lyon had listed, the provost marshal there complained bitterly that he would have to replace their totally inadequate uniforms out of his budget.

While Lieutenant Howell hurried to cover his backside, his bosses at regimental command and at Northwest Service Command in Whitehorse hurried too. They had looked hard for an opportunity to make a point to the dangerously undisciplined Black soldiers of the 97th, to make an example of them. The events at Big Gerstle offered that opportunity and they hastened to seize it. Meanwhile, under arrest in their barracks, Sergeant Heard and his men didn't know it, but they had left the world of soldiers and officers, trucks and weather, details and work, and passed through a rabbit hole into the strict system of military justice that existed in the United States Army in 1942.

CHAPTER TWENTY
The Purpose of a Court Martial

Lieutenant Howell's conference with his temporary commander, Cpt. Jack Doyle, at regimental headquarters rendered the routine issues of the supply depot, the weather, and the trucks irrelevant. Colonel Mitchim wanted to set an example of his recalcitrant Black troops, and he had every officer at headquarters looking for precisely the opportunity that Lieutenant Howell delivered that afternoon. As virtuous examples, these ten young mutineers would do just fine. Doyle sent Howell back to put the men under arrest in quarters with no qualms of conscience, and, informed of the situation, Colonel Mitchim swung into action, equally without qualm. From the perspective of Sergeant Heard and his men, the ludicrous series of events Mitchim had set in motion made no sense, nor would it from the perspective of a modern observer. Sergeant Heard and his men had resisted following orders, and we understand that disobedient soldiers get punished. But Mitchim set the wheels in motion to charge and try the ten men for mutiny—the most serious crime a soldier can commit, a crime on par with murder.[439]

Today, under the Uniform Code of Military Justice (UCMJ), when someone alleges that a soldier violated a law, the system investigates. If the investigation indicates that the soldier did indeed violate a law, the system charges the soldier and brings them to court-martial. In a general court-martial, a panel of officers sits as judges, an attorney prosecutes, and another attorney defends. The entire point? To fairly determine guilt or innocence and impose appropriate punishment. We call it "military justice" today just as they did in 1943. They used terms like "guilt" and "innocence" just as we do today. But the UCMJ didn't exist in 1943. When Northwest Service Command fed Sergeant Heard and his men to the Judge Advocate General (JAG) Corps, it fed them to a very different process. In 1943 JAG Corps officers worked, first and foremost, to help senior commanders do their jobs, to help them control and motivate the soldiers they commanded. Questions of guilt, innocence, and fairness took a backseat when a commander needed to demonstrate the consequences of breaking the rules.

In his book *Court-Martial*, Chris Bray quotes Edmund Morgan, a "law professor and former soldier," to explain: "A court-martial was not a court, but simply an agency of the commanding officer. . . . It was said to be his right hand to help him maintain discipline and was controlled not by law but by his will."[440] And a system controlled not by law but by a commander will turn the army's fundamental racism loose. The belief that Black soldiers required more discipline, and therefore harsher consequences, all too often contaminated the commander's will. Bray also quotes Professor Alice Kaplan: "83 percent of the men executed in Europe, North Africa and the Mediterranean Theaters of Operation were African-Americans, in an Army that was only 8.5 percent Black."[441] Fred Rust, the historian of the white 18th Engineering Regiment, noted that "twelve [white] men were tried by separate courts-martial at various times throughout the mission [the Alaska Highway project]. The most serious charge was willful disobedience to a lawful order and the maximum sentence was six months at hard labor and loss of two thirds of pay for the same period."[442]

The sequence of events that followed the incident at Big Gerstle strikes a modern observer as confused, crazy, and downright surreal. But view the sequence through the eyes of senior commanders in Whitehorse and the eyes of the JAG Corps, and "confused, crazy, and downright surreal" resolves to clarity and common sense.

After Lieutenant Howell confined Sergeant Heard and his men to arrest in quarters, a flood of legal documents surged through the offices of the regiment and Northwest Service Command. Two days after the incident,

Lieutenant Howell handed Colonel Mitchim a signed formal charge of mutiny against each of the ten. Mitchim immediately issued orders appointing Cpt. Jack Doyle to investigate.[443] Doyle commenced taking statements from everyone involved the next day, April 2. He took them from Howell, then Lyon, then the H&S company clerk, the truck drivers, and 1st Sergeant Williams. A thorough investigator, he even took one from the company bugler. Williams, between a rock and a hard place, told his version of the story without mentioning his advice to the men that they would be "damned fools" to get on the designated truck.[444] Doyle finished by interviewing Heard and each of his men, and two days after he started his investigation, he concluded it. He forwarded his stack of statements to Colonel Mitchim and, no surprise, Mitchim topped the stack with his formal recommendation that all ten soldiers be charged with mutiny and tried by general court-martial and forwarded it to General O'Connor.[445] General O'Connor added his formal endorsement to the stack and passed the growing pile of paper to his adjutant general, Lt. Col. J. W. Fraser on April 5. A day later, Colonel Fraser, having "carefully examined and considered [the pile of paper]," added his contribution, formally recommending that the ten be tried by court-martial.[446]

General O'Connor duly signed off on Fraser's recommendation, and Fraser turned the pile of paper over to his trial judge advocate, Cpt. J. Ward Starr, on April 7.[447] Two days earlier, the 97th, knowing in advance the conclusion Colonel Fraser would reach, had moved Sergeant Heard and his men to the Whitehorse stockade.[448] Ironically, they drove the soldiers south on the very highway that, just a few months before, the soldiers had worked so desperately hard to build. The spinning gears of the army and its military justice system had carried Sergeant Heard and his men from the wrong side of the rabbit hole, arrested in quarters at Big Gerstle, to the Whitehorse stockade and the capable hands of Captain Starr in just one week.

The consummate JAG officer, J. Ward Starr grew up in Madison County, Indiana. His father, a pillar of the community and a stalwart in the Republican Party, served for a time as Madison County sheriff. Starr first came to the army as a reserve officer. As a student at Indiana University he was selected to attend a nearby Officer's Training Camp in August 1917 and received a commission in November. Starr graduated from Indiana University in 1919 and returned to Madison County.[449] Starr went to law school. In 1942, when at age forty-seven he registered for the draft in Anderson, Indiana, he lived in a room at the YMCA and practiced law out of an office in the Anderson Bank Building.[450]

The very day Starr registered for the draft in Anderson, April 27, 1942, Sergeant Heard and his men rode the USS *David Branch* through the Gulf of Alaska, nearing snowy Valdez. But, of course, Starr knew nothing of that and almost certainly wouldn't have cared if he had. Later in 1942 Starr merged his careers in politics and law by running for prosecuting attorney, and the good citizens of Madison County elected him. The army, though, had other plans. Starr entered active duty before he could take office.[452] No matter. The army could use his unique skill set in the JAG Corps. If Starr couldn't maneuver through a political labyrinth and prosecute bad guys in Madison County, he could do it in the army. And the army sent him to Northwest Service Command in Whitehorse to do just that as a trial judge advocate.

On April 7 Cpt. J. Ward Starr understood exactly how the military justice system worked. He understood its function as an instrument of control, and he knew exactly how to stage a court-martial with those ends in mind. Captain Starr came to the Whitehorse stockade on April 12 and formally handed each soldier a piece of paper that threatened his very life by charging him with mutiny.[453] And the soldiers met the man who would write, produce, direct, and play a starring role in the rest of their nightmare.

Captain Parsons wrote to Abbie that same day, telling her that "Jim Coleman [Cpt. Jim Coleman of the 97th] was in town for one night last week." Coleman came to "get some cars that were used to bring 10 or 12 old F boys who are to be tried for mutiny."[454] Parsons, a Corps of Engineers officer, didn't understand the function of military justice as an instrument of command any better than we do all these years later. To Abbie he wrote, "I sometimes wonder if its not the officer's fault when these fellows get off the beam. Out of the 12 [sic] I'd bet that only one or maybe two are really bad—the balance are just following." Parsons didn't know it yet, but he planned to take his first step into a more thorough understanding of military justice—way more thorough than he wanted: "I'm going up to see who they are and what's wrong."

Despite Captain Starr's best efforts, the wheels of "justice" could only turn so fast. On May 14, Northwest Service Command issued special orders for a court-martial to begin on May 15 or "as soon thereafter as practicable," listing the officers of the court, reaffirming Starr as trial judge advocate, and naming a defense counsel. But the trial did not begin on May 15. On May 29, 1st Lt. Carl T. Nath, the man Starr had recruited to serve as defense counsel, formally requested a delay "for the purpose of additional preparation of defense, and, appointment of individual counsel."[455] The reason for the delay became clear when, on June 3, Northwest Service Command issued another special order. The accused soldiers had the right to request the services of a defense counsel of their choosing. The new special order, calling for the court-martial to begin on June 5, listed the same court and the same prosecutor. But the listed defense counsel was Cpt. Walter Parsons. The education of Captain Parsons had commenced.

CHAPTER TWENTY-ONE

Setting Up the Performance

In 1943 military justice offered an intricate, complicated, and immensely powerful weapon to its senior commanders—the "General Court-Martial."[457] The complex weapon had two basic moving parts, the "Trial Judge Advocate,"[458] an attorney analogous to the prosecutor in a civilian court, and the "Detail for the Court," a group analogous to judge and jury.[459] Intent on using his weapon to inspire the Black soldiers of the 97th to better discipline, General O'Connor designated Cpt. J. Ward Starr the trial judge advocate in May.[460] In June he designated twelve officers as the detail for the court.[461] Three lieutenant colonels, three majors, five captains, and one second lieutenant—all white men—would array themselves along a table, their uniforms heavy with brass symbols of rank and service, and solemnly, formally determine the fate of the ten Black privates (the army had, of course, already reduced Sergeant Heard to Private Heard[462]). Robert H. Tomlinson, thirty-six, had served on active duty in the army in the mid-1930s as a second lieutenant in the Finance Corps and then stayed on in the reserves. November 1940 found him back on active duty as a captain, still a Finance Department officer. By December 1942 he had risen to lieutenant colonel.[463] Lt. Col. Karl D. Reyer, forty-four, had served in World War I. After the war he earned a PhD and then taught business administration, first at the University of Oklahoma and then at Louisiana State. He returned to the army in September 1940 to serve in the General Staff Corps.[464] Maj. William F. Cleckler, fifty-four, served in World War I,

170   A DIFFERENT RACE

returned to Oklahoma after the war, and went to work for a business in Tulsa. By 1940 he served that firm as a vice president. He had remained active in the reserves, and he returned to active duty as an infantry officer in 1942.[465] Maj. Robert W. Beall served in the Signal Corps. Maj. Mark H. Johnson and Cpt. Edward D. Mazaitis served in the Quartermaster Corps. Captains Farson E. Lynn and Roland M. Anderson served in the infantry. Cpt. Willis W. Finley served in the Corps of Engineers. Cpt. Charles B. Peck and 2nd Lt. Theodore S. Johnson served in the Adjutant General's Department.[466]

When the army assigned these officers to serve on the detail for the court, they received some limited training, along with an overview of the awesome responsibility of their task, and were then sent to perform it. Much like the jury in a civilian court looks to a judge for guidance, these eleven men would look to the twelfth member of the detail, the president and law member, for the same.[467] To that critical role, General O'Connor appointed Lt. Col. Joseph W. Whitney.[468] Whitney, fifty-three, had left high school in Iowa after just two years, knocked about as a civilian for a few years, and then enlisted in the army in 1916. He took well to the army, finding plenty of opportunity there. The army made him a corporal and then a sergeant. In 1918 he served with the Philippine Scouts in Manila, and the army commissioned him a second lieutenant. In the 1920s when a lot of his better-educated fellow officers left the army, looking for better prospects elsewhere, Whitney, with just two years of high school, had no opportunity better than the army. Through the 1920s he served as a captain, first along the border in Nogales, Arizona, then in the Panama Canal Zone.[469] By 1932, when the army appointed Whitney instructor in military science and tactics for the ROTC program at Wilberforce University, it had elevated him to major. In 1937 he transferred to Puerto Rico, and in 1939 the army made him a lieutenant colonel. In 1943 he worked in the Finance Replacement Training Center at Fort Benjamin Harrison in Indiana.[470]

The judge in a civilian court brings knowledge of the law to the courtroom. Colonel Whitney did not bring knowledge of the law to his seat at the center of the table. But as the very definition of a seasoned veteran officer, he brought knowledge that General O'Connor valued far more. Whitney knew the army; he especially understood the court-martial as an instrument of command.[471] He would follow Captain Starr through the legal labyrinth, leading his detail for the court. And the court would make sure General O'Connor got his virtuous example.

A third major player shared the courtroom with Starr and the court. As previously mentioned, Cpt. Walter Parsons served as counsel for the defense. A devout Roman Catholic, Parsons had brought character and a profound sense of right and wrong to the army. He knew little about military justice and nothing about court-martial as an instrument of command, but as a matter of morality, of right and wrong, the proceedings struck him as inappropriate and unfair. If the ten soldiers had resisted obeying orders, they deserved punishment. But mutiny? Possible death sentences? The threat his ten soldiers faced consumed him, and he came to the courtroom determined to mount the best defense possible.[472]

Moreover, as an experienced line officer in the Corps of Engineers, Parsons knew how to command soldiers in the field. And everything about this proceeding flew directly in the face of that experience and knowledge.[473] Parsons knew that soldiers must obey an officer instinctively, not because of abstract army regulations. An officer must maintain and nurture a command presence, a command relationship with his soldiers. Soldiers sometimes resist obeying orders, and Parsons knew the first and most important rule for dealing with that: anticipate resistance to an order and deal with it *before* it happens. Some missions pose a threat to the soldiers who carry them out. A good officer minimizes the threat and includes explanation in his orders.[474]

Parsons also understood the second most important rule for dealing with resistance to orders: if anticipating and explaining fails and soldiers continue to resist, a commander should minimize any potential drama by administering discipline immediately. If a soldier screws up, his commander punishes him. Period. Nothing makes an officer look less like a commander than having to go to his superiors to help him enforce discipline.[475] Parsons knew an offense when he saw one. Clearly these soldiers had balked at following orders. They deserved punishment. But from

Parsons's perspective, knowing nothing of General O'Connor's need to use these men as an example, Lieutenant Howell and his superiors all the way up the line had broken every rule he knew about how to command.[476]

During three days in June 1943, a narrative, a performance, written and directed by Cpt. J. Ward Starr, played out in the makeshift courtroom in Whitehorse. On the morning of June 5, the members of the court took their places at a long table. Colonel Whitney settled in the center. MPs brought ten powerless young Black soldiers from the stockade and positioned them in a row of chairs facing the twelve white officers.[477] Two tables stood between the accused and the court. Polished, confident, experienced attorney Starr had worked on his case since April. He and two assistants took their seats at one of the tables, surrounded by piles of paper and law books. From that table Starr would carefully aim the general's weapon at the young soldiers, and over the next three days he would slowly squeeze the trigger. From the adjacent table, Cpt. Walter Parsons, an experienced line officer with no legal experience and only a few days of preparation, with two assistants he neither knew nor trusted, would try to deflect Starr's bullet.

Parsons watched as Starr set the awesome machine in motion by creating the illusion that "refusal to get on a motor truck" amounted to a capital crime. At 9:40 a.m. he stood to commence an intricate dance of legalities to establish the drama, the awesome responsibility and power of the proceedings. Starr intoned the names of the officers of the court, his own name as trial judge advocate, the names of his assistants, Parsons's name as defense counsel, and the names of Parsons's assistants. He read aloud the special order that established the court-martial. He noted that the charges had been brought by 1st Lt. DeWitt C. Howell and then listed the names of the various officers who had signed off on them.

Then Starr moved to the law: "The evidence that shall be presented in this case are charges that are lodged under the 66th Article of War." He read to those in attendance: "Any person subject to military law who attempts to create or who begins, excites, causes, or joins in any mutiny or sedition in any company, party, post, camp, detachment, guard or other command shall suffer death or such other punishment as a court-martial may direct." He then formally swore in the officers of the court. Colonel Whitney, in his turn, administered the oath to Starr and his assistants and then returned the floor to the master of ceremonies, Captain Starr, who immediately got to the crux of the matter: in a military court as in a civilian court, prosecutors accuse defendants of violating specific laws—the charge; then they detail exactly how the defendants went about doing that—the specification. Starr turned to the accused, directed each man to rise as he called his name and serial number, and formally read the charge that they had violated the 66th Article of War. He then read the specification: "In that Private Willie B. Calhoun, Private James V. Hollingsworth, Private Josh Weaver, Private Robert M. Rucker, Private James M. Heard, Private Sims Bridges, Private Lee I. Ratliff, Private Willie L. Howell, Private Warren H. Lindsey, and Private Eugene Fulks . . . did at Big Gerstle, Alaska, on or about March 29, 1943, cause a mutiny by concertedly refusing to obey the lawful orders of First Lieutenant DeWitt C. Howell, 97th Engineer Regiment, to get upon a motor truck."

Captain Parsons watched his opponent, the masterful trial judge advocate, glide right past the absurdity that "mutiny"—a word redolent of violent rebellion, vile cowardice, even treason—boiled down in this case to "refusing . . . to get upon a motor truck." Starr formally asked, "Does any one of the jointly accused have any special plea or motion to enter at this time?" Parsons declared that the accused did. He moved that the defendants should be tried separately. Starr, of course, objected. A flurry of explanation, of argument and counterargument ensued, punctuated by two long pauses so that the court could confer. But General O'Connor didn't need two courts-martial to make his point. The court denied Parson's motion and directed that the arraignment proceed. The ten privates stood, facing the court, and Captain Starr formally asked each man two questions: "How do you plead to the specifications?" and then "How do you plead to the charge?" Each man answered each question, "Not guilty."

CHAPTER TWENTY-TWO

The Prosecution

Having established the issue as the dire crime of mutiny, Starr turned to the more mundane task that faces every prosecutor in every court: proving the ten had committed it. On the first day Starr called four witnesses—Lt. Robert W. Lyon, Cpt. DeWitt Howell, 1st Sgt. Noah Williams, and Pfc. Clinton Hinson—to present the prosecution's version of the events on March 29 at Big Gerstle. As Starr turned to establishing guilt, Parsons had no choice but to take seriously the process he saw as absurd and to do his best to cast doubt on Starr's points as he and his witnesses made them.[478]

Lieutenant Lyon led off, and, as he would with all of his witnesses, Starr began by leading Lyon through a series of questions to establish his relevance to the proceedings. Lyon shared his name, rank, and his duty in the H&S Company of the 97th as "assistant commanding officer." He confirmed that he worked for Cpt. DeWitt C. Howell, that in March he had worked for then Lt. DeWitt C. Howell, and that on March 29 he had been assigned to and on duty with the H&S Company. He knew the accused, rose, pointed, and named them one at a time—stumbling briefly—but then managed to come up with a name for Eugene Fulks. Finally, Starr turned him loose, and Lyon told his story.

Lyon had planned to take the detachment of ten men to Fairbanks, where he would establish an RSO (regimental supply office). On the evening of the twenty-eighth he discussed his arrangements with his commander, Lieutenant Howell, and on the morning of the twenty-ninth the ten soldiers of Sergeant Heard's squad loaded gasoline, diesel oil, firewood, rations, tents, tent stoves, and other supplies onto three of the four trucks parked outside the orderly room. After lunch, ready to go, Lieutenant Lyon had gone to the barracks and found the men sitting around the stove, talking, clearly not happy. Sergeant Heard asked him about the fourth truck. Did Lyon intend to haul the men to Fairbanks on that one? When Lyon responded yes, Heard objected, arguing that carrying men on a day as cold as this one required a better truck. Lyon went to the orderly room to talk to Lieutenant Howell, who assured him the truck would do fine. Lyon returned to the barracks and told the men to go get on the truck. Not satisfied, Sergeant Heard asked Lyon's permission to speak with Lieutenant Howell. Lyon gave permission and then followed Heard to the orderly room. Lyon heard Lieutenant Howell tell the sergeant to get his men on the truck.

After Heard left the orderly room, the lieutenants heard a commotion outside. Howell went to the door to investigate. The soldiers grouped outside clearly didn't intend to get on the truck, so Howell called them inside, one at a time, ordered each man to get on the truck, and then asked each man if he would comply. One by one the men emphatically refused, and both Howell and Lyon made a list, recording each man's name and serial number. With all the men back outside, Howell sent 1st Sergeant Williams out to line them up alongside the truck. Lyon went out to talk to them, explained mutiny to them, and warned them that the army executed men who committed mutiny.

After a time, Lieutenant Howell emerged from the orderly room. Like Lyon, he explained mutiny and its penalty to the men. He then ordered them to fall in behind him if they intended to board the truck. He gave them ten seconds to obey. During the ten seconds, Lyon heard the men talking, including Sergeant Heard saying that he would go if the others would. But none of the men fell in behind Howell. When the ten seconds elapsed, Howell turned and walked away. Lyon returned to the orderly room for a minute. When he came back out, he found four of the men—Fulks, Hollingsworth, Weaver, and Lindsey–on the truck. Lyon asked them whether they intended to go, and they told him yes. Lyon showed the four how to arrange their equipment in the truck bed, then he climbed off the truck and walked away. He encountered Lieutenant Howell, who told him they weren't sending those four or any other men. Lyon concluded by telling the court that he ordered all of the men to go back to the barracks.

Starr followed up with questions that allowed Lyon to tell the court that he had personally supervised the men of Heard's squad, that Howell had issued clear and understandable orders in a voice loud enough for the men to hear them, and that both he and Howell had explained mutiny and its consequences to the men. Howell left plenty of room between himself and the truck when he issued the ten-second order, and the men stood close enough that they could easily move behind Howell in ten seconds.

Starr asked him about the four men he found on the truck after the ten-second order. Lyon clarified that this had occurred five or ten minutes after the men had initially disobeyed Howell's order. According to Starr, Lyon had also testified that at one time he had seen all ten men on the truck. However, in fact, according to the trial transcript, Lyon had not testified to that. But no one noticed and Lyon confirmed that he had seen all ten men on the truck. Responding to Starr, Lyon told the court that this had occurred around 3:00 p.m., a full hour after Lyon had planned to leave for Fairbanks and forty-five minutes after the men disobeyed Howell's ten-second order.

In cross-examining Lyon, Parsons struggled to find any inconsistency, taking Lyon step by step through his story, but the story didn't change. Parsons questioned Lyon about Howell's order from the door of the orderly room: "You understood that as an order?"

"Yes."

"Could that order have been distinctly heard by every man in that squad?"

"Yes, sir."

Starr took another turn with Lyon and went over the same points. When Parsons came back, he tried a different tack: "Is the witness familiar with the order from the Service Command stating that men are not allowed on open trucks in very cold weather?" The answer to that might have raised some doubts. And, in fact, Service Command had issued such an order, not only forbidding troops from riding in open trucks in such temperatures but from riding in any unheated truck.

Starr leaped to his feet to object. If that issue hadn't come up in direct examination, Parsons couldn't bring it up in cross-examination. Law member Whitney didn't even bother to close and consult with the court detail. He immediately sustained the objection.

Parsons asked Lyon how many men he intended to transport to Fairbanks.

"Sixteen men, sir."

Parsons's next question backfired: "Is it in line to ask where the other six men were?"

"They were there preparing to go on this journey."

Parsons quickly dropped that issue and asked Lyon about the list of names he had taken in the orderly room: "Did you know whether or not Calhoun was on that list?"

"Yes, his name was on the list."

Parsons knew that Willie Calhoun had left the squad after lunch to work out a pay allotment at Personnel, and had not been anywhere near the orderly room. Apparently, Lyon did not know that.

After Lyon escaped the witness stand, Starr brought newly promoted Captain Howell to take his place. He went through a series of routine questions and answers to establish Howell's relevance to the proceeding. Then Starr asked Howell to tell his story.

Howell had ordered a detachment to leave for Fairbanks on the afternoon of March 29. That afternoon, between 1:45 and 2:00 p.m., Sergeant Heard came into Howell's orderly room to question the order to go, arguing that the men couldn't make the trip in the back of that truck in weather this cold. Howell told him they had to go and reminded him that sergeants don't question an officer's orders. Visibly upset, Heard responded that he wouldn't be responsible for his men freezing to death. Howell assured him that wouldn't happen and ordered him to go get his men on the truck. The sergeant left the orderly room.

Thinking the issue was settled, that Heard would follow orders and the truck would depart, Howell went back to work at his desk, but in just a minute he heard noise. He went to investigate and found the whole squad clustered outside the orderly room door. Howell ordered them to get on the truck, but they ignored him, milling around, protesting. He then summoned the men into the orderly room, one at a time, and ordered each man to get on the truck. Some of the men said no. Others said they would go if the others did. Howell added detail to his testimony: "One man in particular, Pfc. Bridges, said he would not go. After he said no, I said to him, 'It seems to me that you have been in enough trouble and that you of all people would want to go.' He said he didn't care how much trouble he was in." The process, Howell said, took about five minutes.

With all the men back outside, Howell said, he sent 1st Sergeant Williams to assemble them. He let them cool their heels for a while and then went outside to confront them one last time. Howell explained mutiny to them, told them that refusing his order amounted to mutiny, and told them that the army executed mutineers. He ordered them to fall in behind him if they intended to get on the truck and gave them ten seconds to obey. Ten seconds passed, but none of the men fell in behind him. Howell left the area and walked away to regimental headquarters. At headquarters he told his commander about the problem and said that these men wouldn't do. They discussed getting a different squad for the trip. Howell finished by telling the court that when he returned to the company area, he ordered 1st Sergeant Williams to send the men back to the barracks under arrest.

Starr's follow-up questions made sure the court understood the key points in Howell's story. Then he turned to the lists of names as the men came into the orderly room: "Did you make a list of their names as they came in?"

"Yes, sir, so did Lieutenant Lyon."

Howell had lost his list, but he now had Lyon's list in hand, and Starr asked him to read the list aloud before he introduced it into evidence. The list Howell read included Private Calhoun. Clearly the fact that Calhoun hadn't actually been there had escaped Trial Judge Advocate Starr and Company Commander Howell just as it had escaped Lieutenant Lyon.

Starr got Howell to repeat, step by step, the sequence of events: He had called the men into the orderly room one at a time, listed their names, and asked them if they planned to get on the truck. They had all said no. He then sent 1st Sergeant Williams to assemble the men. He stood in front of them and explained mutiny and its consequences, he ordered the men to get behind him and gave them ten seconds to comply. None of the men fell in behind him. And he had done all of this to break up what he perceived as a "mass action."

In cross-examination Parsons questioned Howell about H&S Company procedures and then turned to the lists of names: "Did you take all their names?"

"I believe I did."

Parsons asked Howell how many times he ordered the men to get on the truck.

"I ordered them twice."

Parsons turned to the ten-second order. He asked Howell how long he had stayed on the spot after the ten seconds elapsed.

"Just a few seconds."

The answer bothered Parsons, the experienced company commander. The ten-second order flew in the face of everything Parsons knew about commanding soldiers. "You had just given a direct order to ten men to get on a truck, they didn't and you left?" Starr objected, but Colonel Whitney, an even more experienced line officer, wanted to hear the answer. For once he overruled Starr's objection. Parsons tried again: "In ten seconds they didn't get on the truck so you turned around and walked off." Whitney interrupted, pointing out that Parsons needed to rephrase, to ask a question, not make a statement. Parsons rephrased by asking Howell whether he said anything more to the men after he gave the order. Howell responded that he hadn't. When Parsons asked Howell if giving soldiers ten seconds and then walking away was standard practice for an army officer, Starr objected again. This time Whitney sustained the objection.

Parsons asked, "When you were timing the men, giving them ten seconds, did any man make a move towards the truck?"

"I couldn't say because I couldn't see the men then."

That gave Parsons an opening. Howell had just confidently testified that none of the men complied with his order. Now he says he couldn't see them? Starr asked Howell to confirm that he had given three orders. Captain Howell, who had just told Parsons that he gave two, now agreed that he gave three. He gave the first when the men stood outside the orderly room, the second when they came in one at a time, and the third when he gave them ten seconds to fall in behind him at the truck.

Parsons turned to the issue that bothered him most of all: How had someone transformed routine disobedience into the capital crime of mutiny? Parsons asked Howell when he had brought charges and who had suggested doing so. Knowing the true point of the proceedings, Starr and Whitney immediately shut down that line of questioning; Starr objected and Whitney sustained the objection.

The court had questions about the trucks, and Howell told them the trucks were Studebakers—6 × 6 and 6 × 4—and that one of the trucks had a cover. When they asked about the weather that day, Howell remembered wind and a temperature just above zero. According to National Weather Service records, however, on March 29, 1943, Big Gerstle saw a low temperature of thirty-four degrees below zero. Nearby Big Delta set a record of twenty-eight degrees below zero that day.

Starr got Howell to verify that four trucks would have gone on the convoy, that one of those trucks would have carried the men, and that that particular truck had a canvas cover. Howell told Starr that Lieutenant Lyon, four drivers, and two cooks would have accompanied Heard's squad on the convoy.

"Were those two men [the cooks] there?"

"Yes, sir."

"Did they say at any time it was too cold to go?"

"No, sir."

Parsons asked for a clarification: "When Sergeant Heard came into the orderly room the first time, did he ask for another truck?"

"I believe he did."

"Was he talking for himself or for his men?"

"For his men."

The court asked Howell whether he had inspected that truck, and when Howell said he had, they asked about ice on the truck bed. Howell told them that barracks bags covered the ice.

Howell left the stand and Starr brought his third witness, Black 1st Sgt. Noah Williams, to take his place. After the usual questions and answers to establish the relevance of Williams's testimony, Star turned Williams loose to tell his version of the story.

Busy with other things, Williams hadn't paid much attention to the Fairbanks detail until the afternoon when, returning to the company area, he saw the men of Heard's squad squatted in a group next to one of the trucks. Lieutenant Lyon called Williams over, explained that the men had refused to get on the truck, and asked Williams what he should do. Williams told him they should take the men in to the company commander, and Lyon and Williams took Sergeant Heard into the orderly room. Heard saluted Howell, told him the men refused to get on the truck, then went back outside. When Heard returned, he brought the men with him, and Lieutenant Howell summoned them into the orderly room one by one.

Williams essentially told the same story as Lyon and Howell about the men coming into the orderly room and Howell's ten-second order until he got to the men's response to the ten-second order. Williams testified that he saw two men move toward Howell, not behind him but beside him. Williams concluded his story by telling the court that after ten seconds, Howell walked away, and when he returned, he ordered Williams to tell the men to get their things off the truck and put them under arrest in quarters.

Answering Starr's questions, Williams made sure the court knew that Howell had explained mutiny to the men and that the men couldn't have been confused by conflicting orders because only Howell had issued any orders. Starr also had Williams make sure the court knew that Howell had delivered an understandable ten-second order in a voice the men could have heard.

Williams said again that two men had at least moved in response to the ten-second order, and this time he identified them—Hollingsworth and Rucker.

"Did they fall in back of the lieutenant?"

"No, sir, not directly back of him."

"Tell me what your impression was that the men were saying in that squad."

"That they had made up their minds that they were not going to obey the order or ride on the truck."

Starr's questions allowed Williams to tell the court that he had supervised the men in the attached squad for several days, that he had known about the trip for at least a few days, and that he had discussed the plans with Howell. During all that time, he said, no one had complained about the weather.

Parsons followed up on Williams's testimony that Lyon had called him over to ask his advice, but Williams couldn't remember any further detail. Then Parsons took Williams through the sequence of events again. Parsons asked whether during the ten-second order the men could have fit between Howell and the truck, and Williams assured him that they could have. Parsons got him to describe the truck again. Williams told him the truck had a canvas cover but no back flap. And Parsons asked about Willie Calhoun: "Was Calhoun in there?"

"I don't know," Williams responded.

Starr came back to make sure the court knew that Howell had left enough room for the men to get behind him: "Then it is your opinion that all ten men could have fallen in behind him?"

"Yes, sir."

The court took a turn: "How were the men clothed on that day?"

"The usual clothing, sir. Sweaters, fur-lined caps; some had arctics on."

Starr followed up on that issue. He started by questioning Williams about his personal experience operating in cold weather. Williams told the court he had been in the north since April 1942 and had seen lots of bitterly cold weather.

Starr asked, "The men who were detailed to go on this convoy, were they properly clothed to go?"

"Yes, sir."

Parsons took another turn: "Did you inspect the equipment and clothing prior to their departure?"

"No, sir."

Black Pvt. Clinton Hinson, H&S Company mail clerk and assistant to the company clerk, came to the stand at the end of that first day. Hinson told the court that he had worked in the orderly room on the afternoon of March 29. He said he saw Sergeant Heard come in and heard him tell the company commander that his men were refusing to get on the truck to go to Fairbanks. He heard Lieutenant Howell order the sergeant to get the men on the truck. Heard left but returned in a few minutes to report that the men still refused.

Hinson remembered that Howell had called the men in one at a time, ordered them to get on the truck, and asked each if he would comply. Lieutenants Lyon and Howell had written down the men's names as they came into the room. Hinson remembered that Howell had ordered Williams to go out and assemble the men, but Hinson had remained in the orderly room and couldn't describe what happened out by the truck.

When Parsons asked about the lists the lieutenants made, Hinson remembered that Howell had lost his, but they still had Lyon's. Later, after Lieutenant Howell sent the men to their barracks, he sent Hinson to the barracks to make a list of the names and serial numbers of the men under arrest. Hinson didn't know why.

When Private Hinson escaped the witness stand, the court adjourned for the day. The white officers went off to dinner. Military policemen returned the Black soldiers to the stockade

CHAPTER TWENTY-THREE

Adjusting the Story

On the second day of the proceedings, Starr would call Cpl. Willie Stallworth and Tech 5 Howard Graham to the stand. In the first day's testimony, Lieutenant Howell had come across as careless, cold, and distant, not at all concerned for the welfare of his men. Starr needed Stallworth and Graham, two young Black enlisted men in Howell's company, to make Howell appear to be a better officer and a more sympathetic character.[479] Stallworth and Graham reported to Lieutenant Howell. They needed to make him happy. Moreover, they had every incentive to please Starr, the white captain. And Starr had an entire evening to help them prepare their testimony. They came to the courtroom on the morning of June 6 knowing exactly what Starr wanted of them and anxious to provide it.

The court formally reconvened at 9:40 a.m. A burst of formalities from Captain Starr confirmed that everyone and everything occupied the same spaces as when they had adjourned the day before. Then the prosecution called Cpl. Willie Stallworth to the stand. Stallworth explained to the court that on March 29 he had served with the H&S Company of the 97th as a "messenger boy and bugler." He confirmed that "those ten men" had been attached for duty and that he knew of the plan to transport them to Fairbanks. Stallworth started with the story the court had already heard: after lunch that day, Sergeant Heard had come to the orderly room to talk to Lieutenant Howell; argued that the cold made a trip in that truck too dangerous for the men; Howell disagreed and told Heard to get his men on the truck; Heard saluted and left.

From there, however, Stallworth offered a very different order of events. According to him, about thirty minutes later Sergeant Heard "brought his men to the orderly room and . . . said all the men were there." Lieutenant Howell explained to them that they had two or three stops to make and that they could stop whenever they needed to warm up. Moreover, he told them how to keep warm in the truck bed: they should crawl into their sleeping bags. In Stallworth's account Howell didn't make lists of names at that point; instead Howell sent 1st Sergeant Williams to line the men up outside. When Howell went out to talk to them, Stallworth went out too. Howell issued the ten-second order. One man moved in response, but Stallworth didn't know his name. Only after that did Howell call the men into the orderly room one at a time and take their names, ranks, and serial numbers. Then he canceled the trip and ordered Williams to confine the men to the barracks.

The altered sequence didn't bother Starr. His follow-up questions went to the points he most wanted the court to hear: "I believe you said that Lieutenant Howell explained to Sergeant Heard, when he first came in the orderly room, how to put his men on the truck, and to put his men on the truck in their sleeping bags. Is that correct?"

"Yes, sir."

"They were going to stop two or three times. Do you know what the first stop was and how far it was from Big Gerstle?"

"Yes, sir. They were going to stop at Big Delta. From Big Gerstle to Big Delta is some thirty miles."

"About how long would it take to ride to Big Delta?"

"I imagine it would take about one hour and a half."

Starr asked, "Do you know whether or not the men knew that they had planned to stay all night in Big Delta?"

"No, sir. I do not know."

"Was it the general talk in the company that this was the plan?"

"Yes, sir. The general talk in the company was that this convoy was going to stop at Big Delta, which was less than forty miles away."

Messenger boy and bugler Stallworth, indeed the entire enlisted complement of the H&S Company, knew a surprising number of details about the plans for the trip.

When Starr asked Stallworth whether any of the men had said anything in response to the ten-second order, Stallworth remembered, "I heard one man say that he would just as soon die as to go on that truck and freeze to death." He heard another say, "Either we all should go, or if one stays, we all should stay; we all should stick together." He couldn't remember which man had said those things. Starr asked him to identify the man who had moved in response to the ten-second order, and Stallworth stood and pointed at Hollingsworth. Hollingsworth, he said, "walked out of the squad and made an attempt to get behind Lieutenant Howell."

Captain Parsons, when he took his turn with Stallworth, didn't pursue the altered sequence of events. Much more puzzling, he didn't pursue the modifications to the prosecution's story—that is, Lieutenant Howell's assurances that the truck would stop to let the men get warm and his concern that they wrap themselves in their sleeping bags. Parsons wanted to know whether Stallworth knew about Willie Calhoun's absence when Howell took names. Stallworth didn't know. Parsons also wanted to know about the one man who moved in response to the ten-second order: "How close did he fall out to Lieutenant Howell?"

"About one foot from him."

"You saw Lieutenant Howell looking at his watch. He was not looking at the squad when he was looking at his watch. Was it possible that the man could have fallen out behind the lieutenant and he could not have seen him?"

"Yes, it was possible that he could not have seen him."

Starr called Tech 5 Howard Graham to the stand. Graham told the court that he served as a truck driver in the H&S Company, that he backed a truck into the street near the orderly room that day, and that his motor sergeant had dispatched him to drive the truck to carry ten men to Fairbanks.

"Were you going to drive to Fairbanks that day?"

"No, sir, not all the way to Fairbanks that day."

Graham told the court that he saw the men gathered close to the truck.

"Can you tell me what was on this truck?"

"The men's equipment; they had barracks bags, mattress covers; some of them had sleeping bags."

"Was all of their equipment on that truck?"

"Yes."

Starr asked what time they planned to leave, and Graham told him two o'clock. Starr asked how many trucks planned to go. Graham told him four: "The cooks were going with the rations. One truck had rations on it. One had stoves. One had diesel oil and one drum of gasoline." Graham also told the court one of the trucks carried wood.

In fact, that evening, after the men had been returned to quarters under arrest, Graham drove to Big Delta. Starr asked him about that trip. Graham said that Lieutenant Lyon and the cooks went with him, and the trip took "about an hour and a half." It was colder when they started than it was at two o'clock.

"Did you suffer from cold any?"

"No, sir."

"Did any men freeze, or did any man complain of cold?"

"No, sir."

How long, Starr asked, had Graham been driving trucks in "that vicinity"? Graham told him he had done that all winter. Carrying men at fifty below had presented no problem. However, Graham had not transported men through temperatures that cold. Northwest Service Command didn't work or transport soldiers in very cold weather, certainly not at fifty below. Parsons failed to challenge him.

Starr returned to the events at two o'clock and got Graham to describe what he saw. Graham told the court he first saw the ten men when he backed the truck into the company street. Lieutenant Lyon told them to get on the truck, Graham said, and they objected: "They said it was too cold to get on that truck." Lyon and Heard went in to talk to "the company commander." When they emerged from the orderly room, Lyon told the men "what they were doing. . . . It was mutiny and in time of war it was punishable by death." Then Howell came out, talked to them about mutiny, and gave the ten-second order. No one responded to it.

When Captain Parsons cross-examined Graham, he failed to make a dent in Graham's story. Worse, he helped make Starr's point: "Did you hear the lieutenant [Lyon] tell these men that they could stop anytime they wanted to?"

"Yes, sir."

"When did he tell them that?"

"When they said it was too cold to get on the truck."

"Was that before he sent them to see the company commander?"

"Yes, sir."

The court took a turn questioning Graham, asking how he had dressed and how the cold affected him that day. Graham told them he wore pants, a shirt, and a sweater and that he remained outside for up to thirty minutes at a time without needing any extra clothing. The court asked him how cold it was that day. He didn't know. In redirect Starr got Graham to tell the court "it was colder in the morning than the afternoon" and that he thought the men were sufficiently clothed to make the trip.

When Starr called Captain Howell back to the stand, he asked him about the men's clothing that day. Howell could not recall exactly what the men had with them, but they had arctic clothing, and he and Lyon had prepared lists of each man's clothing "on the second day of April 1943." Starr asked him why they had made the lists, and Howell explained, "We were going to send the men to the guardhouse at Whitehorse and I wanted to see what they had." Between March 29 and the day Howell and Lyon made the lists, Howell assured the court, no one had issued the men any additional clothing. Starr asked whether the men had all that was listed in their possession and Howell answered that they did. Starr introduced the lists into evidence and had Howell read each man's list. According to

Howell's lists, Sims Bridges had arctic clothing but only half of a sleeping bag. Fulks's list didn't include a parka, and his footwear consisted of a pair of moccasins. Hollingsworth's list didn't include any footwear. Ratliff had only half a sleeping bag, and his list, like Hollingsworth's, didn't include footwear. Nobody noticed the missing items, not even Parsons. Finally, Starr let Howell confirm for the court that Lieutenant Lyon, in complete charge of the convoy, had the authority to stop and let the men warm up whenever he deemed necessary.

When Captain Parsons rose to cross-examine Howell again, he wanted to know more about the lists Howell and Lyon prepared. Parsons wanted to know whether Howell's company routinely prepared such lists when men left. Howell told him they did. Was Howell the custodian of the men's official records? Howell told him no, he didn't have their Form 32s (the official army form for listing a soldier's clothing and equipment). Starr asked Howell to confirm that the men hadn't acquired any additional clothing between March 29 and the time the officers made their lists. Howell confirmed that.

The court weighed in, wanting to know why commissioned officers had taken on a job normally handled by a supply sergeant—that is, making lists of men's clothing. The supply sergeant, Howell explained, had trouble keeping track of the men's equipment and needed help. Puzzled, the court wanted to know why the supply sergeant had trouble, and Howell explained: "That has a lot to do with the way clothing was issued all last summer; there was lots of mistakes made then."

Parsons cross-examined Howell again and managed to establish that the witness had not inspected the men's clothing for the trip. Howell didn't know whether the men had shoe pacs (waterproof winter combat boots), and he didn't know whether the men's uniforms fit them. He admitted the uniforms were worn and damaged but had not made any lists of worn or damaged clothing. Parsons sat down and Starr took a turn to help Howell explain. His questions and Howell's answers established that the ten men had clothing comparable to that of the other men in H&S.

If Starr wanted to present Howell as a competent and concerned company commander, Parsons effectively established the opposite. He did it by drawing on his own experience as a company commander. Parsons, Howell, and at least some of the officers of the court knew the 97th had a uniform problem and knew about Washburn's scathing January report on the subject. Knowing the inadequacy of his men's uniforms, if Parsons had found it necessary to transport some of them out in an unheated truck in cold weather, he certainly would have inspected their gear and found a way to get what they needed. Clearly that hadn't even occurred to Lieutenant Howell.

After Howell stated very clearly that the clothing the men took to Whitehorse was identical to the clothing they had available for the trip to Fairbanks, Starr excused him and recalled Lieutenant Lyon to the stand. Starr got Lyon to describe preparing lists of clothing with Lieutenant Howell and to confirm that the lists for Whitehorse were identical to what the men had for Fairbanks. He had Lyon read the two lists he had made, and on Lyon's list Willie Howell and Warren Lindsey had no footwear. Lyon also remembered today what he hadn't remembered the day before: the convoy to Fairbanks would have stopped along the way so that the men could warm up. And he explained that to the men, not once but twice.

Parsons then questioned Lyon. Lyon, the supply officer for the H&S Company, couldn't tell him whether the lists of the men's clothing conformed to standard military specifications for arctic equipment. Lyon also didn't know whether Sergeant Heard had a parka. He had not inspected the men's clothing.

The court returned to the ten-second order. Their questions and Lyon's answers confirmed that he had seen four, and later ten, men on the truck, but this had happened only after the ten had already disobeyed Howell's order. The court recessed for lunch and when they returned, the prosecution rested.

CHAPTER TWENTY-FOUR

The Defense

After lunch on the second day, Captain Parsons took over to present his case for the defense. Parsons would defend Willie Calhoun simply by proving the man hadn't been there. But the other nine defendants had clearly resisted following orders. The only way Parsons could defend them was by making them look less guilty, by presenting facts that excused their actions. And, for Parsons, the way Lieutenant Lyon and, especially, Lieutenant Howell had handled them excused a lot.[480] Parsons started with a discrepancy. Corporal Graham's testimony that after Howell's ten-second order none of the men moved toward the truck didn't square with what Graham had told Captain Doyle during Doyle's investigation back in April. According to Doyle's report, Graham had told him all ten men moved toward the truck.

Parsons called Tech 5 Graham as the first witness for the defense: "Did you make the statement that you saw no one approach the lieutenant after the ten seconds was up?"

"Yes, sir."

Parsons offered Doyle's report of his interview with Graham into evidence, and Starr objected, arguing that Parsons was trying to impeach his own witness. The court solved the problem by calling Graham as a witness for the court. And the court asked him, "Did you on or about the 2nd day of April 1943, at the headquarters of the 97th Engineer Regiment append your signature to a statement made by an investigating officer?"

"Yes, sir."

The court turned Graham over to Starr for cross-examination. Starr handed Captain Doyle's account of what he had said to Graham: "Does that conform to the statement that you made verbally to the officer when he made notes?"

"Some of it is wrong. I did not say that they walked over to the truck."

"Did you read these statements before you signed them?"

"No, sir."

Parsons tried to shake that testimony, but he couldn't. The court understood that the investigating officer had not heard the young soldier correctly and had written the mistake into his account of the young soldier's testimony. Offered the chance to read and correct what Doyle wrote, the semi-educated young soldier hadn't taken it; he had simply signed Doyle's account without reading it.

Parsons moved to the next point for the defense. Just a couple of hours earlier, Starr had tried to prove that the ten men had appropriate clothing in good enough repair for the trip to Fairbanks by placing in evidence lists of clothing they had for their trip to the Whitehorse stockade. To do that he had to establish with crystal clarity that the men took to Whitehorse exactly what they would have taken to Fairbanks on March 29. But Parsons knew that they had arrived at the Whitehorse stockade so inadequately clothed that the stockade had to replace *all* of their clothing. Provost Sgt. Henry J. Epperson had complained about the hit to his budget. Clearly, uniforms that were inadequate for the Whitehorse stockade were inadequate for a trip in the back of an unheated truck.

Parsons called Epperson to the stand: "Please tell the court how these men looked and how they were dressed when they came in." Before Epperson could answer, Starr, with breathtaking inconsistency, objected: "How they were dressed when they came to Whitehorse is irrelevant and immaterial." Colonel Whitney, with equally breathtaking inconsistency, sustained the objection. Parsons made the obvious argument: if the clothing lists for Whitehorse represented the clothing the men had at Big Gerstle, then testimony that it was inadequate in Whitehorse amounted to testimony that it was inadequate at Big Gerstle as well. Knowing that the real point of the proceedings was not guilt or innocence but General O'Connor's weapon, his virtuous example, Whitney knew he couldn't allow this. He closed the court so that the members could confer privately. He then reopened the court to rule: "The court considers any testimony concerning the condition of clothing or equipment . . . upon arrival at the post of Whitehorse is irrelevant and immaterial and does not wish to hear any evidence of that nature."

Stymied, Parsons gave up and called defendant Pvt. Warren Lindsey to the stand. When Parsons asked him to tell his story, Lindsey remembered the first part of the day pretty much the way the prosecution had presented it. Notably, from the private's perspective, wherever orders originated, they came to him through his squad leader, Sergeant Heard. At Heard's orders, Lindsey and the men had loaded three trucks in the morning. After lunch, at about one o'clock, Heard directed them to load their gear on the fourth truck. They did so. When it came time to go, Heard left them grouped near the truck and went to the orderly room: "He was going to try and get a better conveyance," Lindsey testified. Heard came back out and told them Lieutenant Howell wanted to see them in the orderly room. Heard went in first and Lindsey followed him. Howell asked Lindsey for his name, rank, and serial number. Lindsey gave that information and went back to the truck. After all the men had gone inside and come back out, Sergeant Heard lined them up. Howell came out and said, "I will give you ten seconds to get on that truck, and if you don't, I will court-martial you and have you put to death." Lindsey, Hollingsworth, Weaver, and Fulks got on the truck. Lieutenant Lyon joined them, and when he asked whether they agreed to go, they answered yes. After Lyon left, Sergeant Heard came to the truck. Angry, Heard said, "You four yellow bitches, I hope you freeze to death." Eventually, Lindsey concluded, Sergeant Heard told them to unload their gear and return to the barracks. They were "confined under arrest of quarters that night."

Parsons asked Lindsey about the squad's interaction with 1st Sergeant Williams, and Lindsey told the court a lot more than Williams had told them: "Before we ever got on the truck, before we went in the orderly room. Sergeant Williams told us, 'You are damn fools to ride on that truck at thirty degrees below zero.' . . . Orders had done come down from Whitehorse from Major Waite or somebody not to ride on trucks when it was that cold." Parsons asked Lindsey how he knew the temperature, and Lindsey told the court he had seen a thermometer reading thirty below zero.

Captain Starr subjected the young private to a lengthy cross-examination, designed less to poke holes in Lindsey's story than to simply make him look unreliable, stupid, and incompetent. His long cross-examination left Lindsey's story essentially intact.

In redirect, Captain Parsons got Lindsey to confirm Calhoun's absence: "When the men lined up behind the truck, how many were there?"

"All but Calhoun."

The court wanted to know whether any men other than the four he had already mentioned got on the truck.

"Just after Lieutenant Lyon got off the truck, two more men got up there. Private Rucker and Private Howell got on the truck."

Starr came back for a short, pointless exchange, and when Lindsey left the stand the court adjourned for the night.

The next morning, June 7, as the trial entered its third and final day, Parsons called Pvt. Josh Weaver to the stand, and Weaver's story matched Lindsey's. Weaver told Parsons that the men had entered the orderly room to talk to Howell one at a time, and when Parsons asked him to point to the men he now remembered being in the line, Weaver pointed to all the defendants except Hollingsworth and Calhoun. Parsons asked about the squad's interaction with 1st Sergeant Williams while they stood in line and waited for Howell to deliver his ten-second order: "Did you hear 1st Sergeant Williams talking to the group of men at any time during that period?"

"Yes. Sergeant Williams said, 'You men stick together, you be a damn fool to ride on that truck as cold as it is below zero.' He said, 'If some of you don't go, none of you go. The law has done passed not to ride on trucks below zero.'"

Lieutenant Howell, Weaver told Parsons and the court, came out and warned them that if they didn't get on the truck within ten seconds, he would have them "court-martialed and put to death." As Howell looked at his watch, Weaver, Hollingsworth, Fulks, and Lindsey walked behind him.

Parsons asked Weaver whether he understood 1st Sergeant Williams's comments as an order not to get on the truck. But Weaver answered no, "I intended to get on."

Starr started his cross-examination by asking Weaver about the temperature that day, and Weaver told him that at about 11:00 a.m. the thermometer on a tree near the Signal Corps building registered thirty degrees below zero. After that, Starr launched a seemingly endless series of questions and answers, going over and over the events of the day. Starr repeated the same questions; Weaver gave him the same answers. When Starr finally sat down, Parsons asked Weaver even more questions about the events of the day. But for reasons known only to him, he also raised a new issue. He wanted to know whether Weaver could read and do numbers. Weaver told him he could "give it a try." That gave Starr an opening.

In his turn, Starr showed Weaver numbers on a piece of paper and asked him to read them. Notably, he asked him to indicate thirty below zero on the paper. When Parsons objected, he walked right into Starr's trap: "If you want to ask the witness, get a thermometer." The prosecution produced a thermometer, and when Starr asked Weaver to point at thirty below zero, Weaver pointed at thirty above. Starr marked the location on the thermometer and introduced it into evidence. On the face of it, Starr made a devastating point, but, in fact, National Weather Service records say the temperature at Big Gerstle that day fell to a low of thirty-four below zero—a record low for that day that still stands.

The court took their turn with Weaver. His answers to most of their questions added no new information, but he confirmed Willie Calhoun's absence during the crucial events that afternoon. And in answer to one member's question, Weaver told the court he didn't know that Lieutenant Lyon planned to haul him to Fairbanks along with the supplies and equipment until Lieutenant Howell told them to get on the truck.

Pvt. James Hollingsworth chose not to testify. Instead he presented a statement giving his version of events. Since he wasn't sworn in as a witness, neither Parsons nor Starr could ask any questions. Hollingsworth told the court he had stepped away to get warm by the fire when Lieutenant Howell called the men individually into the orderly room. When he rejoined the group by the truck, Sergeant Heard sent him in to see Howell, and Hollingsworth told Howell he would go get on the truck. Howell didn't write his name on a list.

When Pvt. Robert Rucker came to the stand as a sworn witness, his story matched the stories told by Lindsey and Weaver. Rucker didn't know they were going to Fairbanks until that afternoon. Rucker didn't like the truck, but he planned to go on it. He and seven others lined up single file outside the orderly room and went in one at a time so that Howell could ask for their names, ranks, and serial numbers. Calhoun and Hollingsworth had not come into the orderly room with the others, but Hollingsworth had gone in a few minutes later.

Parsons asked Rucker about the squad's encounter with 1st Sergeant Williams: "Did Sergeant Williams at any time during that day make any remarks you heard?"

"Yes, sir."

"State when that happened."

"That happened when Sergeant Heard came out of the orderly room for us to line up."

"Was that when Sergeant Williams said something?"

"Yes, sir, he said not to ride on the back of the truck, it was too cold. That an order had passed from Whitehorse not to ride on trucks at thirty degrees below zero."

Parsons asked Rucker about the temperature, and Rucker had seen thirty below on the same thermometer as his squad mates had seen at about the same time. Rucker described the ten-second order and told Parsons that he had moved to get on the truck but waited for Lindsey, Hollingsworth, Weaver, and Fulks to get out of the way and for Lieutenant Lyon to finish showing them how to place their bags. Everybody couldn't pile in at once.

THE DEFENSE   199

Starr cross-examined Rucker exactly as he had cross-examined the other defendants, asking the same questions about that day's events over and over, trying to introduce contradictions or discrepancies. For this witness Parsons tried to object several times. The court overruled him. Rucker told Starr that six men had responded to the ten-second order by moving to get on the truck. Starr wanted to know why it took longer for Rucker to move.

"I was in the back," he said. "I didn't want to walk over them."

The court asked about the men's quarters at Big Gerstle. Rucker told them the squad lived in three different adjacent barracks. Privates Eugene Fulks, Willie Howell, and Willie Calhoun chose to make unsworn statements, and in his statement Calhoun explained why he had not been among the ten. He had been trying to arrange an allotment from his pay; and with Sergeant Heard's permission, he went that afternoon to Personnel to work on that. When he returned, Lieutenant Howell had put the squad, including Calhoun, under arrest in quarters.

That was the end of Parsons's questioning. He had given it his best shot and he rested the defense.

The court now wanted to question Lieutenant Howell, so they called him back to the witness stand. They had taken note of Calhoun's absence from the events of that day, and they wanted to hear the details of Howell's ten-second order and the number of men he had given it to. Howell waffled: "I didn't know the exact number. It was my impression that the whole squad was there. I didn't count them."

Starr took Howell, one last time, through the key points of the prosecution's story. Howell told Starr that he had first given Sergeant Heard an order to get his men on the truck. Second, he had come to the door of the orderly room and ordered all the men to get on the truck. As they came, one at a time, into the orderly room, he gave each man an order to get on the truck. And finally, to the group assembled next to the truck outside the orderly room, he had given the ten-second order. Starr asked about the temperature as well. Captain Howell told him it was ten degrees above zero. Why Howell remembered a temperature so different from the one remembered by the men, we don't know—maybe because he spent most of the day in the orderly room. Regardless, National Weather Service records confirm the men's memories, not Howell's.

Parsons then took a turn to ask Howell whether he had walked away immediately after the ten-second order. Howell told him he turned and looked behind and beside him before he left. Parsons wanted to know whether it was possible that some men had moved.

"I don't believe so," Howell replied.

Parsons also wanted to know what Howell did after he walked away. Howell told him he walked to headquarters to talk to Captain Doyle, the acting regimental commander. Howell asked Doyle whether he could assign a detachment from Company E in place of the men from Company F. He told Doyle he found the Company F men who had "bucked" against obeying his orders unsuitable for the mission. Captain Doyle recommended that he charge the men with mutiny.

Parsons wanted to argue that the men received too many confusing orders: "I am trying to get over to the court that he was giving them so many orders that it had them confused." Over Starr's objection, the court let Parsons proceed: "One time he gave two orders to get on the truck, and the next time to get behind him in ten seconds. The men didn't know whether to get on the truck or to get behind him. You know, some of these men have had only five years schooling and have to think about these things."

Starr asked Howell how he was dressed that day and whether he had any problems with the cold. Howell wore pants, shirt, and a field jacket and hadn't noticed the cold. Starr then addressed the issue of confusion: "In your opinion, was the time elapsing between those three orders so small that a trained soldier would be confused?"

"Not in my mind."

The court wanted to hear again from Lieutenant Lyon, who now struggled to keep his story consistent. The court asked whether Lyon had heard Howell give the men an order more than once. Lyon answered no, just once. The court next asked what time Howell gave the order.

Lyon asked, "The first order?"

The court answered yes.

"About 2:00 p.m."

And when the court asked about the number of men who came into the orderly room and the number of men on his list, Lyon struggled again. He had testified repeatedly that all ten men had been there disobeying orders. Now he said that all ten had been there "as far as I knew." Starr tried to help him: "The only way you have of knowing whether Calhoun was present is that somebody came in the room and gave the name of Calhoun and you wrote it down."

"Correct."

The court wanted to question 1st Sergeant Williams. They had noticed the testimony about the regiment's policy on working in cold weather: "What is the nature of this order that has been published regarding cold weather and working conditions?"

"I do not know the exact wording of it, but it says that men are not to work out of doors if it is more than twenty-five or thirty degrees below zero."

If this time Williams remembered that he had discussed that policy with the men, he still found a way to throw them under the bus. The discussion had occurred not before the ten-second order but only long after, when he confined the men to arrest in quarters. And besides, Williams assured the court, March 29 was a warm day with temperatures above zero. Williams told the court that Howell had given each man an order to get on the truck when he came into the orderly room and that all ten men had come into the room to receive that order. Parsons wanted to know whether anyone had ever read the Articles of War to the ten. Williams didn't know. The court asked whether they had heard the specific Article of War regarding mutiny. Starr guided Williams to an answer. Both Lyon and Howell had explained mutiny and its consequences very clearly. The court recessed for dinner.

CHAPTER TWENTY-FIVE

Summations and the Verdict

General O'Connor needed a dramatic example of what happens to men who resist discipline. Over the past three days in the Whitehorse courtroom, Starr and the court had provided that. When the court reconvened at 7:30 p.m., Starr rose to deliver closing arguments for the prosecution, to bring the drama to its climax.[481] During the last three days, he pontificated, the court "has been trying men for a crime seldom known. Trying men for a deed more dastardly than that at Pearl Harbor. More dastardly because every man who participated in this crime is a citizen of our own country. Mutiny."

Why was this crime so dastardly? "The crime which these men committed strikes at the very foundation of our government, a crime that can be committed only by those who have sworn to uphold and defend at all costs, and with all sacrifices, the Constitution of the United States. It strikes at all that is the United States." To convict them of this dastardly crime the prosecution needed first to prove collective insubordination, meaning the men had plotted together to commit the crime as a group: "The prosecution has shown that these ten men, the accused, joined together, came together, and by their concerted action, accomplished the end and purpose of a common and concerted intent." The only possible defense against the charge of mutiny, Starr explained, would be that it would have been impossible for the ten to obey Howell's order. That rendered the issues raised by the defense—from the temperature that day

203

to the state of their uniforms—irrelevant. The men could have climbed onto the truck, but they didn't. The men committed mutiny.

Starr sat down, and Captain Parsons rose to make a closing argument for the defense. For three days Parsons had operated in a straitjacket as a player in Starr's drama. Parsons knew his defendants had resisted obeying orders. He would have defended them not by trying to prove them innocent but by challenging Starr's drama. Starr and the court had refused to let him do that. Freed at last to speak from his experience and his common sense, Parsons went right to the heart of the matter: Starr and the court had kept the trial scrupulously color-blind. But to Parsons, the fact that his ten defendants were Black men meant everything.

"Now let's consider these men. They are of a different race, . . . a race that has not had the advantages that we have had. . . . You who are from the South know the schooling these men get. You know the attitude of the people in that part of the country." Because Black men don't have the same education and advantages as white men, Parsons said, white officers have an obligation to protect Black soldiers. "I am here to protect these ten men from misunderstanding between them and their officers."

"I was commanding officer of these ten men. Because of that fact I came to know these men. . . . I know how they place their trust in their officers and how also they distrust those who fail to make any effort to help them get along." He turned to the history of the 97th: "Over a period of months, things have been going on and happening and taking place that have caused these men to be suspected of doing something of which they had no intention of doing." He told the court about the regiment and Company F's work on the road, their lack of equipment from the very beginning as they came into Alaska in April: "These men were placed in F Company, a poorly equipped company, . . . poorly equipped with clothing, poorly equipped with cooking utensils, stove, and tents." The men had worked through the summer and into the winter to complete the highway, and "when winter came no housing was there." With Company F these men had wintered at Northway: "They did not object to working or riding in trucks at that time. It was customary. . . . I have seen these men go out and cut wood when it was fifty degrees below zero. . . ." These men, he told the court, had endured. They may have grumbled, but they did their duty. He described Robinson's September reorganization and its impact: "No one had time for saluting, no one had time to correct men, no one had time for court-martials. . . . It became the policy of the regiment that we have no time for this sort of thing."

After Parsons took command of Company F, "Things went along smoothly until we had trouble in the company, like all companies do. Taking this up, I first became familiar with the general policy of the regiment, which was, we do not have time—we do not have time to train these men, to correct these men, but must let them do as they please and get this road built." Parsons had argued to change that policy—to no effect: "And things rocked along and the road was finished and when we got through, the morale of the regiment was on rock bottom." Policies finally changed and the officers of the regiment turned to restoring discipline: "What are the results? These men who have been for months acting just like they acted, doing nothing different than they have been doing for months, were caught short because we had changed our policy."

Parsons made his most fundamental, commonsensical point: "These men did not commit mutiny. Mutiny was not committed—a misunderstanding was committed, or confusion. There was a misunderstanding between officers and men. There was confusion in the minds of both."

Starr had made a big deal of the men's "common intent." Parsons argued that amid all the confusion of the day, "their very actions indicated that if there was any common intent it was to do as they were told." Starr had talked about "collective insubordination." Exasperated, Parsons addressed that issue: "Collective—in that it intended each and every one to commit the same insubordination. They [the prosecution] have not proved that. The witnesses on the stand for the prosecution have indicated that some of these men were willing to go, wanted to go, tried to go."

Parsons asked a rhetorical question: if the men had leaned toward obeying the order, why had they not done so? And he answered it for the court. The men had heard lots of orders on the afternoon of March 29. Their commander gave them an order to get on the truck; minutes later he gave them an order to come into the orderly room: "A good soldier knows the last one [order] is the one to obey," Parsons said. Howell, he pointed out, had testified that in the orderly room he ordered the men individually to get on the truck. He didn't go out to see if they did so. He sat inside to tend to some other work. Then he sent his first sergeant, Sergeant Williams, to line them up. So were they supposed to get on the truck or were they supposed to line up?

The temperature at Big Gerstle that day hadn't made it impossible for the men to get on or ride in the truck, but the temperature explained, and justified, Heard's actions. After living with temperatures of fifty and seventy below, men feel warmer at thirty below. They can walk around without arctic gear for short periods of time. But "try riding on a truck and riding down the highway when it is thirty degrees below. You will find that it is quite a bit different." Out of concern for his men, Heard had tried to persuade Howell to get a better truck. The army had trained the men to follow Heard's lead and they did so instinctively.

Parsons found Howell's attitude and actions that day incomprehensible. Howell had testified that after he witnessed an act of mutiny, an act so heinous that it was punishable by death, he left and went to see the acting company commander. And what did they discuss? Mutiny? No. They discussed getting a squad of men from Company E to replace the squad from Company F. If the company commander hadn't yet realized that the dastardly crime of mutiny was afoot, how could anybody expect the ten men to realize it? "Gentlemen, it is the just opinion of the defense counsel that these men are guilty of nothing more serious than minor disobedience, and if you take into consideration that that sort of thing had been going on for months, you can't find them guilty of that."

Parsons had very effectively called the entire process, Starr's drama, into question. To defend the drama, Starr rose one last time to attack Parsons and his common sense as irrelevant. First he simply dismissed the entire complex issue of "a different race": "Gentlemen—mutiny is never known by race nor color nor creed." And then he turned directly to attack Parsons: "The defense made assertions not based on evidence, others misstating the evidence." Moreover, "The defense admitted the accused to be guilty . . . of disobedience. Disobedience is the very essence of mutiny."

Parsons had talked about trust, the trust that soldiers—especially Black soldiers—must have in their commanders. The army, Starr argued, requires its soldiers to have that trust whether their commanders have earned it or not: "By all their military training and discipline, by their sworn oath, it was the duty of these men to trust their superior officer. They repudiated their oath and trust." Parsons had talked about inadequate equipment. Starr dismissed that too: "It does not require equipment for a soldier to obey a command of a superior officer. . . . One can get on a truck naked." Parsons had recounted the history of the regiment, the lack of discipline. Starr found that irrelevant: "I am not interested in what has gone before. . . . If the defense counsel knows mutiny has been committed on other occasions, then it is his duty to bring it before this court in the proper manner." Parsons had criticized Captain Howell as a commander. Offended, Starr argued that "above all things [Howell] had discharged his duty." Starr closed by making the real point of the proceedings crystal clear to the court: "It is you who must decide whether or not the rebuilding of this regiment, . . . the maintenance of proper military order, shall be a part of our personal contribution to the war effort."

The court closed to consider their verdict. If Parsons had accomplished nothing else, he had proved that Willie B. Calhoun hadn't even been there. The court judged Calhoun not guilty. The other nine men, however, they declared guilty. The court closed again to consider sentences. When they returned, they threw the book at Heard and sentenced him "to be confined at hard labor . . . for twenty years." They threw the book almost as hard at Sims Bridges, sentencing him "to be confined at hard labor . . . for eighteen years." They found Lee Ira Ratliff a bit less culpable and sentenced him "to be confined at hard labor . . . for twelve years." Rucker and Howell had moved toward the truck. The court sentenced them "to be confined at hard labor . . . for five years." Hollingsworth, Weaver, Lindy, and Fulks had climbed onto the truck. The court sentenced them "to be confined at hard labor . . . for three years."

Captain Parsons never understood what had rolled over him and the ten defendants in that courtroom. He wrote to Abbie on June 8:

> Pal I've been several days without sending you a letter but if you only knew what I had to do you would understand. My 10 colored boys wanted me to defend them at the trial. If you think defending 10 men in a mutiny trial for which, if found guilty, they could be put to death is not a big job, just try it sometime. I'm so tired, and have been each day, that I don't know what to do. It's harder than hard work. . . . I'm not sure if I won or lost. One fellow was set free. Four given three years, two given five years, one given twelve years, one given 18 years and the Sgt was given twenty years. Pal it made me sick when I heard them read the sentences but when most of the men came over and thanked me for trying I felt better. . . . The fellow who was set free wanted to stay with me always, so he said when it was over. . . . I hope I never have such a duty again in my whole life.

CHAPTER TWENTY-SIX

Punishment

Around midnight, Captain Parsons returned to his quarters to write to his wife. Starr and the detail for the court returned to their quarters. Guards marched the ten soldiers back to confinement in the stockade. Willie B. Calhoun would leave them in the morning to return to duty, albeit with a different outfit. The rest would remain in limbo in the stockade. Starr had done his job. And, of course, the rumor mill had done its job. Every man in the 97th understood the awful fate that awaited men who disobeyed their white officers. Starr and the detail for the court returned to their normal duties, and Captain Parsons returned to the Canol Project.

Up in Alaska the soldiers of Mitchim's reorganized, reinvigorated, disciplined 97th had got on with their duties too. In June, with the dreadful winter behind them, the soldiers of the regiment worked through the long, warm days of an Alaska summer, slept through the short nights in relative comfort, and ate well, at least by army standards. Still scattered, the regiment now centered on Livengood, still preparing to build General Somervell's road to Nome.[483] Company F spent June in North Fork, repairing the road to Circle, Alaska, rebuilding bridges.[484]

Meanwhile, the nine mutineers waited in the stockade, marking time, nervously waiting for the staff judge advocate to review the trial and its results. The review, they understood, couldn't make their situation worse. But would it make it better? They waited for fifty days.

Lt. Col. Samuel B. Lippitt sent his conclusions to General O'Connor on July 27, and on July 30 General O'Connor issued formal orders implementing them.[485] First, he argued, "The fact that some half an hour later all of the accused showed evidence of willingness to comply with the order and an apparent abandonment of insubordinate purpose is not a defense but may be considered in extenuation."[486] Second, he concluded, "The evidence . . . is legally sufficient to support the findings of guilty as to accused Rucker, Heard, Bridges, Ratliff, and Howell."[487] Third, though, "The evidence in this case is legally insufficient to prove beyond a reasonable doubt that accused Lindsey, Hollingsworth, Weaver, and Fulks were guilty."[488] As to appropriate sentences for the five guilty men, Lippitt offered this: "There was evidence of persistent, concerted disobedience. . . . There was, however, an absence of violence or serious disturbance."[489] Further, Lippitt argued, "Considering all the facts of the case, including extenuating circumstances, the age of each of the accused, and no previous convictions, it is believed that the interests of justice and discipline will be best served by reducing the sentence to confinement at hard labor and the dishonorable discharge be suspended."[490] He reduced Rucker's and Howell's confinement from five years to two years, Ratliff's from twelve years to five years, Heard's from twenty years to ten years, and Bridges' from eighteen years to ten years.

In July Lindsey, Hollingsworth, Weaver, and Fulks returned to duty with new outfits. Military police escorted the five remaining mutineers from the Whitehorse stockade down the long road to the lower forty-eight, all the way to Battle Creek, Michigan. Through August, as they traveled south under guard and in shackles, enduring the stares and curiosity of strangers along the way, the soldiers of their regiment in Alaska had less work to do, and their officers introduced training programs to keep them busy. In early September, when the guards delivered the five mutineers to Fort Custer,[491] the other soldiers of their regiment finally left Alaska.[492] The regiment convoyed back down the long road to Valdez that they had struggled so hard to build. And on September 6 they boarded a troopship at the Valdez dock. On September 11 they docked in Seattle.[493]

At Fort Custer the mutineers landed in the middle of an army experiment. The army had invented a new kind of confinement facility. They didn't call the facility at Fort Custer a prison or a jail, they called it a Service Command Rehabilitation Center.[494] Through 1942, with war raging all over the world, the army consumed soldiers at an appalling and rapidly increasing rate. It occurred to some senior officers that a lot of their prisoners might not be irredeemably bad—just misguided. It might make sense to establish confinement facilities dedicated to retraining these men and returning them to active duty. In an address delivered in November 1944, Robert B. Patterson, undersecretary of war, explained:

> We must give a square deal in the Army to the prisoner. At the same time we must make it our concern that in the prison system we do not provide an attractive shelter for delinquents and shirkers. . . . In military prisons there are offenders of the same sort as in civil prisons, and there are also those who have committed offenses of a purely military character. . . . The man guilty of absence without leave, willful disobedience or some other military offense is in a class by himself from a criminal viewpoint. . . .
>
> We who are charged with operating military prisons will always bear in mind our duty to the Army as a whole. . . . In time of war it is expected to exercise a strong deterrent effect on potential offenders, a policy that points in the direction of treatment that is not lax and a conservative trend on clemency. At the same time it must do what it can to salvage manpower, a policy that points toward rehabilitation and a liberal trend on clemency.[495]

Make no mistake, being rehabilitated, the men served time in a confinement facility, surrounded by "guns, guards and barbed wire."[496] High fences "bedded in concrete" surrounded the area, and "machine guns, tower guards and sentries combine[d] to enforce detention."[497]

The mutineers, like all new prisoners, underwent physical and mental examinations, and then guards put them in quarantine in "barracks inside a double stockade."[498] After two weeks they were moved into the general population. There they started each day with an hour of drill and then spent eight hours working at various menial jobs, such as hauling coal and collecting garbage.[499] Through the days they moved from place to place in the stockade, closely supervised by cautious guards. As the prisoners labored, they had access to "tools, knives, and other articles which would assist in an escape or can be used to injure other persons."[500] Guards carefully searched prisoners working on such details before they moved them elsewhere.[501]

The five mutineers behaved themselves, but not all of their fellow prisoners did. And the guards dealt with that. Lt. Col. G. C. Greenwalt, commanding officer of the center at Camp Phillips, Kansas, described that process for a reporter: "Many pretty tough offenders . . . are softened up by punishment. . . . There are jail cells, with bars of stout wood—and there are several stand-up cells, such narrow cubbyholes that the inmates cannot possibly sit down, or even kneel. He stays there eight hours a day while other inmates are working outdoors for the same length of time."[502]

Prisoners moved through the system, either rehabilitating or not. From the "general population section," prisoners that officers deemed incorrigible donned uniforms marked with a "P" and served time as third-class prisoners "segregated under close guard."[503] The mutineers and other prisoners who rehabilitated satisfactorily moved from the general population section to a "training battalion in which he drills like other soldiers."[504] Those who continued to please the officers and psychiatrists with their behavior ultimately moved to an Honor Company: "The Honor Companies, composed of those who are nearly ready for restoration, are in practically a duty status, in some instances living in quarters outside the fence or wall."[505]

In October 1943, a month after the mutineers arrived at Fort Custer, the Service Command Rehabilitation Center moved from Fort Custer to Fort Sheridan, Illinois.[506] While the mutineers moved under guard on a special train to Fort Sheridan, the soldiers of their regiment lived at Camp Sutton in North Carolina.[507] In March 1944 the center moved again, to Camp Phillips, Kansas.[508] By the time the mutineers moved to Kansas, their regiment had moved to Camp Stoneman in California and then taken a ship for Australia.[509]

James Heard and his four fellow mutineers finally made it to an Honor Company at Camp Phillips. And in May and June 1944 all five privates returned to normal duty—not, of course, with the 97th.[510] In May and June the soldiers of the 97th were maintaining airfields and roads at hot and humid Milne Bay in New Guinea. They would do that through the end of the war in 1945.

* * *

Somebody had committed a crime. Either nine young soldiers at Big Gerstle, Alaska, committed mutiny, or the United States Army in Whitehorse, Yukon, perpetrated a gross miscarriage of justice. But honorable men, the young Black soldiers and the army officers who punished them, did their duty as they saw it. The perpetrator who committed the crime wasn't a person. The perpetrator was the invisible hand that "ruled . . . the lives of all the colored people in . . . the entire South."[511] The invisible hand came to the army with Black soldiers. The army had fed ten young soldiers into the system of military justice to set an example. Having made its point, the army returned them all to normal duty. But they returned to normal duty and, ultimately, they left the army to live their lives in a world still ruled by the invisible hand and regulated by the laws of Jim Crow.

EPILOGUE
The Ten Back in the World

The story of the 97th Engineers came to a climax in that Whitehorse courtroom. The ten defendants could have been just about any ten soldiers in the regiment. The commanders needed an example and these lucky ten happened to provide the first opportunity. If they ultimately left the army to live their lives in a world still ruled by the invisible hand, so did the rest of the Black soldiers in the 97th.

For that reason, we want very much to share with you what happened to the ten back in that world. Be aware, though, that most of what happened to them has disappeared in the mist of history. We can only share what we know.

Willie Calhoun (born May 5, 1923, in Elberton, Georgia; died February 3, 1974, in Atlanta, Georgia)
The court acquitted Willie Calhoun, and in December 1943 he reported for duty with the 258th Signal Corps Construction Company.[512] But Willie had other problems. He had been injured while working on the Alaska Highway in 1942. His bulldozer took down a tree and it fell back on him. Doctors in the field hospital checked him out, but in the end his injuries healed, and they sent him back to duty.[513]

In 1943, confined with the others in the Whitehorse stockade, waiting for their courts-martial, Willie fell and sprained his left knee. This time the Whitehorse station hospital treated him and returned him to the stockade.[514]

By the time Willie got to the 258th, he complained of chronic pain and couldn't perform his duties. In October the company sent him to the Station Hospital in Whitehorse. Doctors there couldn't find any physical reason for the pain, so in January they sent him down to Baxter General Hospital in Seattle for observation by a psychiatrist.

The psychiatrists at Baxter diagnosed a psychoneurosis that interfered with performance of military service.[515] The army gave him an honorable medical discharge in August 1944, and he returned to Georgia, not to Elberton but to Cobb, Georgia, near Atlanta. He fell in love with Annie Lee. They got married and raised a daughter, Lovie Mae, and two sons, Willie Ben Calhoun III and John Calhoun.[516]

Nobody knows why, but on February 3, 1974, a neighbor, Eddie Lee Edmonson, shot Willie in the forehead and chest and killed him.[517]

Eugene Fulks (born December 11, 1918, in Vicksburg, Mississippi; died October 5, 1946, in Vicksburg, Mississippi)

The Staff Judge Advocate reversed Eugene's conviction and returned him to active duty with the 1349th Engineer General Service Regiment. At the end of the war, that regiment was in Europe. Eugene returned to the United States on the USS *John Ericsson* and disembarked in New York on October 9, 1945. Discharged that same month, he returned to his home in Vicksburg, Mississippi.[518]

A year later, in October 1946, his wife, Mamie, stabbed him to death. In 1947 a court sentenced her to four years in prison for manslaughter.[519]

Warren H. Lindsey (born December 31, 1920, in Wilson, North Carolina; died February 4, 2001, in Elm City, North Carolina)

The Staff Judge Advocate reversed Warren Lindsey's conviction and returned him to active duty. The army discharged him as a Tech 5 on October 14, 1945, and he returned to North Carolina to spend the rest of his life in Wilson County. In 1947 he married Lillie R. Chisel, and the couple raised three daughters—Ethel, Jacqueline, and Cynthia.[520]

Warren worked as an auto body repair man for Lee Motor Company in Wilson, North Carolina. When he died at eighty at the Wilson Medical Center in 2001, he left behind his wife and daughters, two grandchildren, and one great grandchild.[521]

Josh Weaver (born March 10, 1919, in Keysville, Georgia; died February 13, 1987, in Savannah, Georgia)
The Staff Judge Advocate reversed Josh Weaver's conviction, and Josh returned to active duty with the Transportation Corps. He served with them in the Southwest Pacific for the duration of the war. When the army discharged him in December 1945, he returned to Georgia and settled in the Richmond Hill neighborhood of Savannah.[522]

He married Fannie Inez Taffar and worked for a steel products company until his retirement in 1986. He died in February 1987 in Savannah, Georgia.[523]

James V. Hollingsworth (born January 13, 1913, in Jeffersonville, Georgia; died October 30, 1994, in Milledgeville, Georgia)
The Staff Judge Advocate reversed James Hollingsworth's conviction and returned him to active duty. Discharged from the army at Fort McPherson, Georgia, on January 12, 1945, James returned to his Georgia home and spent his life there. He never married. James died in Milledgeville, Georgia, about thirty miles from his Jeffersonville birthplace on October 30, 1994, at the age of eighty-one.[524]

Robert M. Rucker (born August 8, 1918, in Cannon, Georgia; died May 10, 1972, in Lexington, North Carolina)
Robert Rucker served his time in the Service Command Rehabilitation System. He made it to the Honor Company at Camp Phillips, Kansas. The system released him to active duty in late spring 1944. Assigned to the Transportation Corps, he served with the 866th Port Company until the army discharged him at the end of January 1946.[525]

Robert returned to Rowan County, North Carolina, and made his life there. In 1949 he married Essie Simpson. Robert and Essie raised three sons, Donald, Ronald, and Raymond, along with two daughters, Diane and Margaret. He worked first for several area dairies as a bottle washer and, according to his daughter, Diane, learned to make fantastic ice cream at home. Ultimately he found a job at Fiber Industries. Robert attended the Shady Grove Baptist Church, where he served on the board of deacons and was the church janitor and a member of the men's choir.[526]

Robert didn't talk to his family about his army service. His daughter had no idea her dad had served in Alaska or had anything to do with the Alaska Highway. She certainly had no idea he had been convicted of mutiny.[527] Diane remembered a loving, hardworking father who took good care of his kids. He teased them sometimes, maintained discipline, and worked long hours to make sure he could offer them every opportunity in life.[528]

Robert died tragically in an auto accident in May 1972. Teaching Margaret, his youngest daughter, to drive he traveled a country road with her at the wheel. Something mechanically locked in the steering mechanism, and the car left the road and hit a tree. Robert was fifty-three.[529]

Willie Howell (born September 15, 1918, in Madison, Georgia; died July 25, 2003, in Madison, Georgia)
Willie served his time with the Service Command Rehabilitation Center, made the Honor Company at Camp Phillips, and returned to duty in June 1944. When the army discharged Willie, he returned to Madison County, Georgia.[530] He married Bertha Carter and the couple raised four sons and one daughter.

Willie died on July 25, 2003, at eighty-four in Madison County, Georgia. [531]

Lee Ira Ratliff (born August 16, 1922, in Norwood, North Carolina; died January 15, 2000, in Salisbury, North Carolina)
Lee Ratliff served his time, made the Honor Company at Camp Phillips, and returned to duty in June 1944. When the army discharged him in January 1946, Lee returned to Stanly County, North Carolina.[532] He married

Zola Mae Watkins, and together they had three sons and one daughter. A second daughter died in infancy.[533] Ultimately Zola Mae found a man she liked better and left Lee to raise the kids by himself.[534]

Lee spent most of his life after the army working at Young Manufacturing. Les Young, the plant manager during the years Lee worked there, explained to us that Lee had an especially responsible and difficult job. The plant manufactured furniture, much of it out of custom-built plywood. Lee mixed the dyes and spread the glue for coloring and sealing the plywood.[535]

He died in Salisbury, North Carolina, in January 2000, at the age of seventy-seven.[536]

James M. Heard (born March 24, 1923, in Elberton, Georgia; died June 24, 1972, location unknown)
James served his time in the rehabilitation system, made it to the Honor Company at Camp Phillips, and in May 1944 he returned to duty at the Army Service Forces Training Center at Fort Leonard Wood, Missouri. The army discharged James in January 1946,[537] and he all but disappeared from history.

James did not return to his hometown, Elberton, Georgia, and figuring out where he spent the rest of his life is part detective work and part just plain guessing.

His mother, Sally Mae Heard, lived in Elberton in 1941 when James enlisted. But at some point she moved to Prince Georges County in Maryland. She passed away there in 1987. His father, Chris Heard, moved at some point to Spotsylvania County in Virginia. He passed away there in 1974. We know that four of James's siblings passed away in Prince Georges County and one more in nearby Washington, DC.[538]

Pretty clearly the Heard Family joined what Isabel Wilkerson called "America's Great Migration." It's reasonable to guess that James joined it with them.

James died at age forty-nine on June 24, 1972. The record doesn't tell us where.[539]

Sims Bridges (born November 22, 1919, in Prichard, Alabama; died [date and location unknown])
Sims served his time in the Service Command Rehabilitation System, made the Honor Company and returned to active duty in May 1944. Assigned to a new unit, Sims wound up in France where, in April 1945, he assaulted a female civilian. The army restored his conviction, and this time he did not do his time in the rehabilitation system. He served it in the United States Penitentiary in Terre Haute, Indiana.[540]

Upon his release, civilian Sims Bridges relocated to upstate New York, where in October 1952 a civilian court convicted him of third-degree burglary, five counts of second-degree forgery, and five counts of petty larceny. The civilian court sent him to the New York state prison at Attica.[541]

When Attica released him, Sims moved to Rochester, New York. In 1958 he burglarized Model Beauty Shop, and the court sent him back to Attica for another five years.[542] Five years later, out of prison and back in New York, he broke into Albert's Jewelers. Still another court sent him back to Attica—for ten years this time.[543]

Sims left Attica on parole in January 1971, just in time to miss the famous Attica Prison Riot in 1971.[544]

When Attica released him on parole, Sims apparently retired from his life of crime. He completed his parole at age sixty-two on October 22, 1981, and disappeared from history.[545]

NOTES

Abbreviations Used

CB   Camp Blanding Museum, Starke, Florida
NA   National Archives, College Park, Maryland
NPRC   National Personnel Records Center, St. Louis, Missouri
UAF   University of Alaska, Fairbanks, Alaska, Elmer E. Rasmuson Library, Alaska and Polar Regions Collections
USACE   US Army Corps of Engineers Office of History, Humphreys Engineer Center, Fort Belvoir, Virginia
USAHEC US Army Heritage and Education Center, Carlisle, Pennsylvania

Notes to Chapter

[1] Isabel Wilkerson, *The Warmth of Other Suns* (New York: Vintage Books, 2010), 31.
[2] Ibid., 54.
[3] Ibid., 87.
[4] Ibid., 50.
[5] Sharyn Kane and Richard Keeton, *In Those Days: African American Life Near the Savannah River* (Atlanta: National Park Service, Southeast Region, 1994), 62. This is an oral history funded by US Army Corps of Engineers.
[6] Aurolyn Melba Hamm, *Elbert County, Georgia, Black American Series* (Charleston, SC: Arcadia Publishing, 2005), 9, 11.
[7] Maj. Perry V. Wagley, psychiatrist with Army Medical Corps, Sixth Service Command, Rehabilitation Center, Department of Psychiatry and Sociology, Fort Custer, Michigan, report on James Heard, October 7–8, 1943, National Personnel Records Center, St. Louis, Missouri (hereafter, NPRC).
[8] Robert A. Margo, *Race and Schooling in the South: A Review of the Evidence* (Chicago: University of Chicago Press, 1990), chapter 2, 6–32.
[9] Wagley, psychiatric report on Heard, NPRC.
[10] Ibid.

[11] Ibid.
[12] Ibid.
[13] C. Eric Lincoln and Lawrence H. Mamiya, *The Black Church in the African American Experience* (Durham, NC: Duke University Press, 1990), 7–9, 11, 93, 171.
[14] Ibid.
[15] Wagley, psychiatric report on Willie Howell, NPRC.
[16] Military service record of Willie Calhoun, with medical, social, and personal history, NPRC; Ancestry.com.
[17] Ibid.
[18] Ibid.
[19] Military records of James Hollingsworth and Josh Weaver, Ancestry.com and Fold3.com.
[20] Les Young interview by author, summer 2018. Young was an employer of Lee I. Ratliff, Company F, 97th Regiment, as well as a resident and local historian of Norwood, North Carolina.
[21] Charles W. McKinney Jr., *Greater Freedom: The Evolution of the Civil Rights Struggle in Wilson, North Carolina* (Lanham, MD: University Press of America, 2010), 3–5.
[22] Warren H. Lindsey in 1930 census, Ancestry.com.
[23] McKinney, *Greater Freedom*, xvii; "Lynching in Wilson First since 1921," *Statesville Record* (North Carolina), August 21, 1930.
[24] Sims Bridges military records, marriage records, census 1920, 1930, Ancestry.com. and Fold3.com.
[25] Eugene Fulks in 1930 census, Ancestry.com.
[26] Albert Dorsey Jr., "A Mississippi Burning: Examining the Lynching of Lloyd Clay and the Encumbering of Black Progress in Mississippi during the Progressive Area," unpublished manuscript, Spring 2009, 6–7, 22, 44.
[27] Diana Rucker Phifer interview by author, 2018. Diana is the daughter of Tech 5 Robert M. Rucker, Company F, 97th Engineers.
[28] Wilkerson, *Warmth of Other Suns*, 8.

Notes to Chapter 2
[29] Military enlistment records, Ancestry.com.
[30] "Engineer Training Center at Fort Belvoir, Virginia," *Moberly Monitor-Index* (Alabama), August 16, 1941; "Selectee Completes 13-Weeks Training, Ft. Belvoir, VA," Gettysburg Times (Pennsylvania), October 16, 1941.
[31] Camp Blanding Company C Roster and Reports, 1941, Walter Parsons private collection (hereafter, WP).
[32] Walter Parsons III interview by author, May 4, 2018. Parsons is the son of Cpt. Walter Parsons, Company F Commander, 97th Engineers. Ancestry.com, Newspapers.com.

[33] Jim Ashton, "Camp Blanding: The War Years," http://www.30thinfantry.org/blanding_history.shtml; "Camp Blanding or BUST—This Is the Way It Was," *Bradford County Telegraph* (Florida), August 10, 1989.
[34] Ashton, "Camp Blanding."
[35] "Race Troops Get First Hutments," *Pittsburgh Courier*, March 21, 1942.
[36] "Whites Trained as Combat Soldiers; Negroes Do Work," *Pittsburgh Courier*, July 19, 1941; "Soldiers Learn at Blanding: Five Teachers Instruct in Elementary English, Typewriting and Shorthand," *Pittsburgh Courier*, July 19, 1941; "Movies Real Feature at Blanding," *Pittsburgh Courier*, July 19, 1941; "Opens Center for Soldiers in Jaxon," *Pittsburgh Courier*, March 21, 1942.
[37] P. L. Prattis, "Prejudiced Southern Officers Should Be Transferred," *Pittsburgh Courier*, July 19, 1941.
[38] Transfer Order No. 35, 97th Engineer Battalion, Camp Blanding to Eglin Air Field, Special Order #35, November 7, 1941, WP.
[39] "Training Troops Begin Ahead of Schedule: Improvements at Eglin Field to Cost $1,500,000," *Miami News*, October 6, 1940; "Negro Troops at Eglin Field," *Panama City News-Herald* (Florida), November 19, 1941; "Negro Troops Reach Eglin: Will Be Used to Clear Ranges," *Pensacola News Journal* (Florida), November 20, 1941.
[40] Walter Parsons, History of the 97th Engineer General Service Regiment, 1941–1943, WP.
[41] Fred Bryson interview by author, summer 2018. Bryson is the son of Tech 5 Thad Bryson, Company B, 97th Engineer Regiment.

Notes to Chapter 3
[42] Philip H. Godsell, *Romance of the Alaska Highway* (Toronto: Ryerson Press, 1944), 115.
[43] "President Franklin D. Roosevelt approved joint proposal of the Secretary of War, Navy and Interior for Construction of the Highway," War Department memorandum, February 11, 1942, US Army Corps of Engineers Office of History (hereafter, USACE); "To Chief of Engineers from Secretary of War. Orders to construct a pioneer type road from Fort St. John, Canada to Big Delta, Alaska," War Department memorandum, February 14, 1942, USACE.
[44] Notes on Cabinet Meeting, January 16, 1942, with Secretary Ickes, Secretary of War, Navy, and Interior to confer and agree on necessity for the highway and which route, US Army Heritage and Education Center (hereafter, USAHEC).
[45] Notes on Meeting, February 2, 1942, by Secretary Stimson, Secretary of War with Secretary of Navy, Secretary of Interior and Brig. General R. W. Crawford to obtain route surveys and survey of possible equipment for road building and report back in one week, USAHEC.
[46] John Schmidt, *This Was No DYXNH Picnic* (Alberta, Canada: Gorman and Gorman, 1991), 39, 40, 41.

47 Memorandum, February 14, 1942, to Chief of Engineers from Secretary of War. Orders to construct a pioneer type road from Fort St. John, Canada to Big Delta, Alaska. Signed by Joseph L. Clark, Adjutant General by order of Sec. of War, USACE.
48 Letter from Gen. Clarence L. Sturdevant to Gen. Simon Buckner, Commander of United States Forces in Alaska. Answering Letter from Gen. Simon Buckner to Gen. Clarence L. Sturdevant April 1942, Lael Morgan Collection, University of Alaska, Fairbanks (hereafter, UAF).
49 Memoirs of Gen. William M. Hoge, Interview by Lt. Col. George Robertson, January 4–15 and April 16–17, USAHEC; Bill Gifford, "The Great Black North," *Washington City Paper* 13, no. 40, October 8, 1993.
50 Heath Twichell, *Northwest Epic: The Building of the Alaska Highway* (New York: St. Martin's Press, 1992), 56, 118–22, 130–31; Gen. William M. Hoge, Progress Reports Whitehorse Sector of Alaska Highway, 1942, Walter E. Mason file, Jane Haigh Collection, UAF.
51 Twichell, *Northwest Epic*, 140, 145.
52 Elizabeth A. Tower, *Icebound Empire: Industry and Politics on the Last Frontier, 1898–1938*, 3rd ed. (Louisville, KY: Old Stone Press, 2015), 3–6, 11, 27–30; Geoffrey Bleakley, "History of the Valdez Trail," Wrangell–St. Elias National Park and Preserve, Alaska, April 14, 2015, 1–13, https://www.nps.gov/wrst/learn/historyculture/history-of-the-valdez-trail.htm.
53 Tower, *Icebound Empire*, 4–5, 8; Cpt. W. R. Abercrombie, *Alaska, 1899, Copper River Exploring Expedition* (Washington, DC: US Government Printing Office, 1900), 37–47; Valdez Museum, "History of Valdez," https://www.valdezmuseum.org/exhibits/resources/short-history-of-valdez/, 2–4.
54 Abercrombie, *Alaska 1899*, 16–26, 65–79; Bleakley, "History of the Valdez Trail," 1–13.
55 Alaska History and Cultural Studies, *Alaska Heritage*, "Road Transportation," ch. 4–10, 1–6, www.akhistorycourse.org/alaskas-heritage/chapter-4-10-road-transportation; Alaska History and Cultural Studies, "Overland Exploration," *Alaska Heritage*, ch. 4–2, 1–3, www.akhistorycourse.org/americas-territory/alaskas-heritage/chapter-4-2-overland-exploration; Mead and Hunt and Cultural Resource Consultants, "Alaska Roads Historic Overview: Applied Historic Context of Alaska Roads," Alaska Department of Transportation and Public Facilities, February 2014, www.dnr.alaska.gov/parks/oha/publications/akroadshistoricoverview.pdf.
56 Regimental Progress Reports, 97th Regiment, 1942–1943, WP.
57 Ibid.
58 Theodore A. Huntley, "Construction of the Alaska Highway: First Year 1942 and Second Year 1943," Public Roads Administration Reports, Washington, DC, September 1945, 114, https://themilepost.com/wp-content/uploads/2017/03/Construction_of_the_Alaska_Highway.pdf; letter from Gen. C. L. Sturdevant to Mr. Thomas H. MacDonald, PRA commissioner, authorizing construction of sections of the Canadian-Alaska

Military Highway, March 31, 1942, Record Group (hereafter RG) 160, National Archives (hereafter, NA).

59 Letter from Sturdevant to MacDonald, March 31, 1942; emphasis added.

Notes to Chapter 4

60 Gen. Dwight D. Eisenhower, War Department, Adjutant General's Office, Washington, SECRET. IMMEDIATE ACTION, "Movement Orders, Shipment #1864," Code Designation 1864-A, Unit: 97th Engineers (GS), Strength: 46 Officers, 1 Warrant Officer, 1259 Enlisted Men, April 7, 1942, WP.

61 Letters, 1942–1943, WP.

62 Eisenhower, "Movement Orders, Shipment #1864"; Letters, 1942–1943, WP.

63 Regimental Progress Reports, 97th Regiment, 1942–1943, WP.

64 Letters, 1942–1943, WP.

65 John Virtue, *The Black Soldiers Who Built the Alaska Highway: A History of Four U.S. Army Regiments in the North, 1942–1943* (Jefferson, NC: McFarland, 2013), 65.

66 Letters, 1942–1943, WP.

67 Robert Tessmer, "Life Aboard a Troop Transport," George C. Marshall Foundation, 2010, https://www.marshallfoundation.org/100th-infantry/wp-content/uploads/sites/27/2014/06/LIFE_ABOARD_A_TROOP_TRANSPORT.pdf.

68 Company B Morning Reports, April 22, 1942, 97th Engineers, 1942–1943, NPRC.

69 Tessmer, "Life Aboard a Troop Transport."

70 Ibid.

71 Valdez City, "Old Town Walking Tour," Valdez Museum, www.valdezmuseum.org/category/education/old-town-walking-tour; "A City Dock and Cannery Dock," *Fairbanks Daily News-Miner*, March 28, 1955.

72 Valdez 1940 census, Ancestry.com.

73 Murray Lundberg, "The Evolution of the Richardson Highway," Explore North, 2011, www.explorenorth.com/library/yafeatures/bl-richardson.htm.

74 Valdez 1940 census, Ancestry.com.

75 Lundberg, "Evolution of the Richardson Highway"; Mead and Hunt, "Alaska Roads Historic Overview."

76 Valdez Museum, "History of Valdez," 2–4, archived at Valdez Museum and Archives, www.valdezmuseum.org/exhibits/resources/short-history-of-valdez.

77 "Bus Company Makes First Highway Trip," *Fairbanks Daily News-Miner*, May 31, 1941.

78 "Alaska Motor Stages, Ride with an Experienced Operator in Modern Deluxe Busses with Individual Reclining Chairs," *Fairbanks Daily News-Miner*, May 8, 1940; "Motor Stage to Start," *Fairbanks Daily News-Miner*, May 8 ,1940; "Buses Leave Fairbanks on Way to Valdez," *Fairbanks Daily News-Miner*, June 12, 1941.

79 Helen Hegener, *Alaskan Roadhouses: Shelter, Meals, and Lodging along Alaska's Early Roads and Trails* (Wasilla, AK: Northern Light Media Publishers, 2015), 117, 188–95;

Kenneth L. Marsh, *The Trail: The Story of the Historic Valdez–Fairbanks Trail That Opened Alaska's Vast Interior* (Trapper Creek, AK: Trapper Creek Museum, 2008), 237.
[80] "Patients Are Brought Here from Valdez," *Alaska Miner*, January 21, 1941.
[81] "And I Saw A New World. Chapter 5," *West Schuylkill Press and Pine Grove Herald* (Pennsylvania), July 31, 1939.
[82] Valdez Museum, "History of Valdez."
[83] Coleen Mielke, "El Nathan Children's Home, Lazy Mountain Children's Home, and Victory Bible Camp," www.freepages.rootsweb.com/~coleen/genealogy/alaskaorphanage.html.
[84] Valdez 1940 census, Ancestry.com.
[85] Margaret Keenan Harrais, *Remembering Old Valdez*, ch. 20, p. 12, unpublished autobiography, n.d., archived at Alaska Room, Valdez Alaska Library.

Notes to Chapter 5
[86] John Virtue, *The Black Soldiers Who Built the Alaska Highway: A History of Four U.S. Army Regiments in the North, 1942–1943* (Jefferson, NC: McFarland, 2013), 65.
[87] Valdez City, "Old Town Walking Tour"; Morning Reports, 97th Engineers, 1942–1943, Companies A, B, C, D, E, F, H&S, April 29, 1942, NPRC; Regimental Progress Reports, 97th Regiment, 1942–1943, WP.
[88] Photo of Valdez dock, Valdez Alaska, https://discovervaldezalaska.org/discover/history/.
[89] Virtue, *Black Soldiers*, 39.
[90] Interview with Hayward Oubre, "Building the Alaska Highway," *American Experience*," PBS television documentary, directed by Tracy Heather Strain, narrated by Joe Morton, season 17, episode 4, aired December 31, 2004.
[91] Virtue, *Black Soldiers*, 96.
[92] Regimental Progress Reports, 97th Regiment, 1942–1943, WP.
[93] Ibid.
[94] Ibid.
[95] Regimental Progress Reports, 97th Regiment, 1942–1943, WP; Company E Morning Reports April 30, 1942, NPRC.
[96] Regimental Progress Reports, 97th Regiment, 1942–1943, WP.; Company B Morning Reports May 3, 1942, NPRC.
[97] Regimental Progress Reports, 97th Regiment, 1942–1943, WP; Company C and F Morning Reports, May 10, 1942, NPRC.
[98] Regimental Progress Reports, 97th Regiment, 1942–1943, WP.
[99] Ibid.
[100] Hoge, Progress Reports Whitehorse Sector, 1942–1943, NPRC; Regimental Progress Reports, 97th Regiment, 1942–1943, WP.
[101] Virtue, *Black Soldiers*, 129.

[102] Ibid., 78.
[103] Interview with William Griggs, "Building the Alaska Highway," *American Experience*, PBS television documentary, directed by Tracy Heather Strain, narrated by Joe Morton, season 17, episode 4, aired December 31, 2004.
[104] Virtue, *Black Soldiers*, 78.
[105] Ibid., 96.
[106] H. Milton Duesenberg, *Alaska Highway Expeditionary Force: A Roadbuilder's Story* (Clear Lake, IA: H & M Industries Ltd., 1995), 24.

Notes to Chapter 6
[107] Messages, 97th Regiment, 1942–1943, WP.
[108] Letters, 1942–1943, WP.
[109] Messages, 97th Regiment, 1942–1943, WP; Letters, 1942–1943, WP.
[110] Letters, 1942–1943, WP.
[111] Messages, 97th Regiment, 1942–1943, WP.
[112] Letters, 1942–1943, WP.
[113] Messages, 97th Regiment, 1942–1943, WP.
[114] Ibid.
[115] Stan Cohen, *The Forgotten War: A Pictorial History of World War II in Alaska and Northwestern Canada* (Madison, WI: Pictorial Histories Publishing, 1981), 86; Morning Report 297th Infantry Battalion, Alaska Army National Guard, 1942, NPRC.
[116] Regimental Progress Reports, 97th Regiment, 1942–1943, WP; Company B Morning Report, May 1942, listing thirty men from Company B to act as stevedores, NPRC.
[117] William E. Griggs, *The World War II Black Regiment That Built the Alaska Military Highway*, ed. Philip J. Merrill (Jackson: University Press of Mississippi), 27 (photo).
[118] Regimental Progress Reports, 97th Regiment, 1942–1943, WP.
[119] Ibid.; Company D Morning Report, NPRC.
[120] Regimental Progress Reports, 97th Regiment, 1942–1943, WP; Company C Morning Reports, NPRC.
[121] Pvt. Major Banks, US Army service records, 1942, NPRC.
[122] Ibid.
[123] Medical Department, Department of the Army, *Preventive Medicine in World War II*, vol. 5: *Communicable Diseases Transmitted through Contact or by Unknown Means*, ed. Lt. Gen. Leonard D. Heaton and Col. John Boyd Coates Jr. (Washington, DC: Office of the Surgeon General, 1960), 411, 419–29.
[124] Letters, 1942–1943, WP.
[125] Duesenberg, Alaska Highway Expeditionary Force, 56, 58, 64.
[126] Ibid.
[127] Ibid.
[128] Ibid.

Notes to Chapter 7

[129] Huntley, "Construction of the Alaska Highway," 114.
[130] "CAA Airfield Construction Firms Listed," *Fairbanks Daily News-Miner*, June 13, 1941; "Big Delta CAA Field Is Started," July 5, 1941; Duesenberg, *Alaska Highway Expeditionary Force*, 32.
[131] Duesenberg, Alaska Highway Expeditionary Force, 32.
[132] Ibid, 35.
[133] Ibid, 34.
[134] Ibid., 33.
[135] Letters, 1942–1943, WP.
[136] Duesenberg, *Alaska Highway Expeditionary Force*, 38.
[137] Ibid.
[138] Ibid., 39.
[139] Ibid.
[140] Ibid., 40.
[141] Ibid.
[142] Ibid., 41.
[143] Ibid.
[144] Ibid.
[145] Ibid., 42.
[146] Ibid., 43.
[147] Ibid., 45.

Notes to Chapter 8

[148] Regimental Progress Reports, 97th Regiment, 1942–1943, WP.
[149] Abercrombie, *Alaska 1899*, 108.
[150] Regimental Progress Reports, 97th Regiment, 1942–1943, WP.
[151] Ibid.
[152] Ibid.
[153] Paul C. Raso interview by author, summer 2018. Raso is the son of 1st Lt. Joseph E. Raso, Company B Commander.
[154] Virtue, *Black Soldiers*, 99; Duesenberg, *Alaska Highway Expeditionary Force*, 25.
[155] Caterpillar Tractor Company, Fifty Years on Tracks (Chicago: Photopress, 1954), 62.
[156] Blanche D. Coll, Jean E. Keith, and Herbert H. Rosenthal, *United States Army in World War II: The Corps of Engineers: Troops and Equipment* (Washington, DC: US Government Printing Office, 1958), 237.
[157] Doyle, Jack F. Captain, Historical Information of 97th Engineer Regiment, Military Report NA943, NPRC.
[158] Regimental Progress Reports, 97th Regiment, 1942–1943, WP.
[159] Ibid.

[160] Ibid.
[161] Ibid.
[162] Duesenberg, *Alaska Highway Expeditionary Force*, 45.
[163] Letters, 1942–1943, WP.
[164] Twichell, Northwest Epic, 211.
[165] Memorandum from Lt. Col. Lionel E. Robinson to all officers, "*OUR JOB*," July 31, 1942, WP. Robinson had just taken command of the regiment and been promoted to lieutenant colonel when he wrote this memo to all of his officers.
[166] Ibid.
[167] Letters, 1942–1943, WP.
[168] Ibid.
[169] Ibid.
[170] Regimental Progress Reports, 97th Regiment, 1942–1943, WP.
[171] Company F Morning Report, June 16, 1942, NPRC.
[172] Michael Morton, interview by author, 2018. Morton is the son of Sgt. James "Jimmie" D. Morton, H&S Company, 97th Engineers; Job in the 97th from Ancestry.com; Military Register for James D. Morton, Ancestry.com.
[173] Griggs, *World War II Black Regiment*, 111, 112.
[174] Letters, 1942–1943, WP.
[175] Morning Report, 297th Infantry Battalion, June 30, 1942, NPRC.
[176] Letters, 1942–1943, WP.
[177] Medical Department, Department of the Army, *Preventive Medicine in World War II*, 5: 419.
[178] Pvt. Major Banks, US Army Service Records, NPRC.
[179] Ibid.; Letters, 1942–1943, WP.
[180] Letters, 1942–1943, WP.

Notes to Chapter 9
[181] Brian Garfield, *The Thousand Mile War: World War II in Alaska and the Aleutians* (Fairbanks: University of Alaska Press, 1969), 31–40, 119, 273.
[182] Ibid.
[183] Letters, 1942–1943, WP.
[184] Ibid.
[185] Ibid.
[186] Duesenberg, *Alaska Highway Expeditionary Force*, 35.
[187] Ibid., 36.
[188] Ibid., 43.
[189] Ibid., 63.
[190] Ibid., 35–36, 38.
[191] Ibid.

Notes to Chapter 10

[192] Duesenberg, *Alaska Highway Expeditionary Force*, 45.
[193] Ibid., 46.
[194] Ibid., 49.
[195] Regimental Progress Reports, 97th Regiment, 1942–1943, WP.
[196] Duesenberg, *Alaska Highway Expeditionary Force*, 48.
[197] Ibid., 49.
[198] Letters, 1942–1943, WP.
[199] Duesenberg, *Alaska Highway Expeditionary Force*, 56; "Gus Ostermann" biography, December 19, 2003, 1–3, Sioux Biographies (Sioux County, Iowa), http://iagenweb.org/boards/sioux/biographies/index.cgi?read=41598, site maintained by Linda Ziemann.
[200] Ostermann biography, 1–3.
[201] Duesenberg, *Alaska Highway Expeditionary Force*, 59.
[202] Ibid.
[203] Ibid.
[204] Ibid., 55, 61.
[205] Twichell, *Northwest Epic*, 211.
[206] Regimental Progress Reports, 97th Regiment, 1942–1943, WP.
[207] Letters, 1942–1943, WP.
[208] Ibid.
[209] Ibid.
[210] Ibid.
[211] Ibid.
[212] Ibid.

Notes to Chapter 11

[213] Abercrombie, *Alaska 1899*, 58–69, 97–98.
[214] Ibid.
[215] Ibid.
[216] Twichell, *Northwest Epic*, 211.
[217] Virtue, *Black Soldiers*, 99.
[218] Ibid.
[219] Ibid., 98.
[220] Duesenberg, *Alaska Highway Expeditionary Force*, 56.
[221] Ibid.
[222] Ibid., 69.
[223] Ibid.
[224] Ibid.
[225] Ibid.
[226] Ibid.

227 Ibid., 73.
228 Ibid.

Notes to Chapter 12
229 Regimental Progress Reports, 97th Regiment, 1942–1943, WP.
230 H&S Company Morning Reports Aug. 2, 1942, NPRC.
231 Letters, 1942–1943, WP.
232 Regimental Progress Reports, 97th Regiment, 1942–1943, WP.
233 Twichell, *Northwest Epic*, 211.
234 Robinson, *OUR JOB*, memo, WP.
235 Ibid.
236 Regimental Progress Reports, 97th Regiment, 1942–1943, WP.
237 Letters, 1942–1943, WP.
238 Ibid.
239 Donna Blasor-Bernhardt, *Pioneer Road: Recollections of the Pioneers Who Built the Alaska Highway* (Las Vegas: ArcheBooks Publishing, 2004), 28.
240 Letters, 1942–1943, WP.
241 Ibid.
242 Fred Bryson interview by author, summer 2018.
243 Company B Morning Report November 9, 1942, NPRC.
244 Letters, 1942–1943, WP.
245 Regimental Progress Reports, 97th Regiment, 1942–1943, WP.

Notes to Chapter 13
246 Twichell, *Northwest Epic*, 212.
247 Duesenberg, *Alaska Highway Expeditionary Force*, 75.
248 Letters, 1942–1943, WP.
249 Blasor-Bernhardt, *Pioneer Road*, 62–63.
250 Willie Calhoun, US Army Service Records, NPRC.
251 Rosters, 97th Engineers MOS (Military Occupational Specialty) code, NPRC.
252 Ibid.
253 Robert Platt Boyd, *Me and Company "C"* (n.p.: self-published, 1992), 93–94; Jon Krakauer, "Ice, Mosquitoes, and Muskeg: Building the Road to Alaska," *Smithsonian* 23, no. 4 (1992): 102.
254 Boyd, *Me and Company "C,"* 93–94; Duesenberg, *Alaska Highway Expeditionary Force*, 78–79.
255 Boyd, *Me and Company "C,"* 95–96.
256 Ibid.
257 Duesenberg, *Alaska Highway Expeditionary Force*, 64.
258 Ibid., 24–25, 28.

[259] Bryson interview.
[260] Regimental Progress Report, 97th Regiment, 1942–1943, WP.
[261] Ibid.
[262] Ibid.; Blasor-Bernhardt, *Pioneer Road*, 20.
[263] Twichell, *Northwest Epic*, 212.
[264] Regimental Progress Reports, 97th Regiment, 1942–1943, WP.
[265] Blasor-Bernhardt, *Pioneer Road*, 189.
[266] Letters, 1942–1943, WP.
[267] Hoge, Progress Report Whitehorse Sector, August 1942, NPRC.
[268] Duesenberg, *Alaska Highway Expeditionary Force*, 85.
[269] Ibid., 73.
[270] Ibid.
[271] Ibid., 85.
[272] Ibid., 64.
[273] Ibid.
[274] Ibid., 85.
[275] Ibid., 75.
[276] Ibid., 76.
[277] Hoge, Progress Reports Whitehorse Sector, June, July 1942, NPRC.
[278] Tech 5 Fred Rust, *Role of the Eighteenth Engineer Regiment: April 1942 to January 1943*, p. 20, USACE. Rust was the historian for the 18th Regiment.
[279] Ibid., 26.
[280] Ibid., 27.
[281] Ibid., 31.

Notes to Chapter 14
[282] Regimental Progress Reports, 97th Regiment, 1942–1943, WP.
[283] Ibid.
[284] History of the 97th Engineer General Service Regiment, 1941–1943, WP.
[285] Letters, 1942–1943, WP.
[286] Ibid.
[287] Regimental Progress Reports, 97th Regiment, 1942–1943; Letters, 1942–1943, WP.
[288] Messages, 97th Regiment, 1942–1943, WP.
[289] Ibid.
[290] Ibid.
[291] Ibid.
[292] Ibid.
[293] Ibid.
[294] Ibid.
[295] Ibid.
[296] Regimental Progress Reports, 97th Regiment, 1942–1943, WP.

297 Ibid.
298 Duesenberg, *Alaska Highway Expeditionary Force*, 75.
299 Company E Morning Report August 30, 1942, NPRC.
300 History of the 97th Engineer General Service Regiment, 1941–1943, WP; Regimental Progress Reports, 97th Regiment, 1942–1943, WP.
301 Duesenberg, *Alaska Highway Expeditionary Force*, 52.
302 Ibid., 56.
303 Letters, 1942–1943, WP.
304 Griggs, *World War II Black Regiment*, 47.
305 Letters, 1942–1943, WP.
306 Hoge, Progress Reports Whitehorse Sector, 1942–1943, NPRC.
307 Rust, *Role of the Eighteenth Engineer Regiment*, 9–10.
308 Regimental Progress Reports, 97th Regiment, 1942–1943, WP.
309 Griggs, *World War II Black Regiment*, 41.
310 Letters, 1942–1943, WP.
311 Ibid.
312 Blasor-Bernhardt, *Pioneer Road*, 28.
313 Virtue, *Black Soldiers*, 152.
314 Bryson interview.
315 Rust, *Role of the Eighteenth Engineer Regiment*, 18.
316 Duesenberg, *Alaska Highway Expeditionary Force*, 63.
317 Ibid.
318 Ibid., 77.

Notes to Chapter 15

319 Rust, Role of the Eighteenth Engineer Regiment, 9; Duesenberg, *Alaska Highway Expeditionary Force*, 99.
320 Virtue, *Black Soldiers*, 100.
321 Company B Morning Reports, 97th Engineers, NPRC.
322 Ibid.
323 Letters, 1942–1943, WP.
324 Virtue, *Black Soldiers*, 100.
325 Rust, *Role of the Eighteenth Engineer Regiment*, 31.
326 Ibid.
327 Ibid.
328 Ibid., 33.
329 Ibid.
330 Ibid.
331 Letters, 1942–1943, WP.
332 Ibid.
333 Twichell, *Northwest Epic*, 205.

Notes to Chapter 16

[334] Rust, *Role of the Eighteenth Engineer Regiment*, 33, 36.
[335] Ibid., 33.
[336] Ibid., 35.
[337] Ibid., 34.
[338] Regimental Progress Reports, 97th Regiment, 1942–1943, WP.
[339] Letters, 1942–1943, WP.
[340] Twichell, *Northwest Epic*, 213.
[341] Ibid., 210.
[342] Gen. William M. Hoge, 1974 interview by Lt. Col. George Robertson, USAHEC.
[343] Twichell, *Northwest Epic*, 213.
[344] Ibid., 208, 209.
[345] Ibid.
[346] Rust, *Role of the Eighteenth Engineer Regiment*, 36.
[347] *General Court-Martial Record of Trial in Matter of the United States vs. Willie B. Calhoun, et al.*, CM-239042, appointed by the Commanding General of the Northwest Service Command at Whitehorse, Yukon Territory, Canada, June 5–7, 1943, Military Report, summation by Walter Parsons, NPRC.
[348] Huntley, "Construction of the Alaska Highway," 25; and Twichell, Northwest Epic, 213.
[349] Duesenberg, *Alaska Highway Expeditionary Force*, 90.
[350] Letters, 1942–1943, WP.
[351] Rust, *Role of the Eighteenth Engineer Regiment*, 39, 40.
[352] Regimental Progress Reports, 97th Regiment, 1942–1943, WP.
[353] Letters, 1942–1943, WP.
[354] Ibid.; Letter from Lt. Col. E. G. Paules to Brig. Gen. C. L. Sturdevant, December 10, 1942, 1–5, USACE.
[355] Regimental Progress Reports, 97th Regiment, 1942–1943, WP.
[356] Letters, 1942–1943, WP.
[357] Duesenberg, *Alaska Highway Expeditionary Force*, 95.
[358] Twichell, *Northwest Epic*, 214.
[359] Ibid, 214–15.
[360] Ibid., 215.
[361] Ibid.
[362] Rust, *Role of the Eighteenth Engineer Regiment*, 43–44.
[363] Twichell, *Northwest Epic*, 216.
[364] Ibid., 218.
[365] Ibid.
[366] Ibid., 219. According to Susan Butler, Roosevelt thought he had a lot more road than really existed. In the spring of 1943, he actually proposed to drive, with Canadian prime minister Mackenzie King, up the Alaska Highway to a meeting with Josef

Stalin at the Bering Strait! Thank God they relocated to Yalta. Butler, *Roosevelt and Stalin: Portrait of a Partnership* (New York: Vintage Books, 2015), 48–49.

[367] Duesenberg, *Alaska Highway Expeditionary Force*, 92.

Notes to Chapter 17

[368] Duesenberg, *Alaska Highway Expeditionary Force*, 101.
[369] Virtue, *Black Soldiers*, 108.
[370] Duesenberg, *Alaska Highway Expeditionary Force*, 114.
[371] Twichell, *Northwest Epic*, 219.
[372] Letters, 1942–1943, WP.
[373] Ibid.
[374] Ibid.
[375] Regimental Progress Reports, 97th Regiment, 1942–1943, WP.
[376] Ibid.
[377] Ibid.
[378] Blasor-Bernhardt, *Pioneer Road*, 22.
[379] Ibid.
[380] Ibid., 24.
[381] Ibid.
[382] Regimental Progress Reports, 97th Regiment, 1942–1943, WP.
[383] Letters, 97th Regiment, 1942–1943, WP.
[384] Messages, 97th Regiment, 1942–1943, WP.
[385] Ibid.
[386] Walter Parsons III interview.
[387] Messages, 97th Regiment, 1942–1943, WP.
[388] H. Bradford Washburn, Investigation and Report of 97th Engineer Regiment Clothing, January 26, 1943, RG 160, NA; Homer D. Angell, Remarks in the House of Representatives, Thursday, March 2, 1944, in Stan Cohen, *ALCAN and CANOL: A Pictorial History of two Great World War II Construction Projects* (Missoula, MO: Pictorial Histories Publishing, 1992), 172–74.
[389] Washburn report.
[390] Ibid.
[391] Judy Ferguson, "Bert and Mary's, Part Two, Guide and Gold Rush Trail Baker," *Delta Wind*, February 19, 2015, 1–5, https://www.deltawindonline.com/features/ferguson/bert-and-mary-s-part-two-guide-and-gold-rush-trail-baker/article_d7274d24-b7a1-11e4-b7a8-3388e6d1b560.html.
[392] Lyons interview.
[393] Duesenberg, *Alaska Highway Expeditionary Force*, 101.
[394] Capt. Richard L. Neuberger, "Yukon Adventure," *Saturday Evening Post*, February 19, 1944, 19–21, 101–102.
[395] Ibid., 102.

[396] Ibid.
[397] Griggs, *World War II Black Regiment*, 11.
[398] Interview with William Griggs, "Building the Alaska Highway," *American Experience*, PBS.
[399] Virtue, *Black Soldiers*, 131.
[400] Lyons interview.
[401] Letter from T/5 Joseph S. Smith to Lt. Lucian C. Lytz, 97th Engineers, Company E, February 6, 1943, Lael Morgan Collection, UAF.
[402] Letters, 1942–1943, WP.

Notes to Chapter 18
[403] Regimental Progress Reports, 97th Regiment, 1942–1943, WP; Twichell, *Northwest Epic*, 223.
[404] Letters, 1942–1943, WP.
[405] Ibid.
[406] Twichell, *Northwest Epic*, 149, 156, 159–61; Stan Cohen, *ALCAN and CANOL: A Pictorial History of Two Great World War II Construction Projects* (Missoula, MO: Pictorial Histories Publishing, 1992), 217.
[407] Letters, 1942–1943, WP.
[408] Twichell, *Northwest Epic*, 236; Regimental Progress Reports, 97th Regiment, 1942–1943, WP.
[409] Letters, 1942–1943, WP.
[410] "Where Is the Coldest Place in Alaska?," *Deep Cold: Alaska Weather and Climate*, November 30, 2012, https://ak-wx.blogspot.com/2912.
[411] Debbie Heral et al., *Delta Junction: At the End of the Alaska Highway*, "Temperatures 19, The Wind 34," brochure prepared for the Delta Chamber of Commerce by TriDelta Inc., printed by Dragon Press, 1992.
[412] Twichell, *Northwest Epic*, 227–28.
[413] Ibid., 224–25.
[414] Huntley, *Construction of the Alaska Highway*, 30.
[415] Company F Morning Reports and Letters, 1942–1943, WP.
[416] Letters, 1942–1943, WP.
[417] Regimental Progress Reports, 97th Regiment, 1942–1943, WP.
[418] Military Register for Charles F. Mitchim, Ancestry.com, Fold 3.com.
[419] Ulysses Lee, *The Employment of Negro Troops* (Washington, DC: US Government Printing Office, 1963), 44.
[420] Ibid., 45.
[421] Virtue, *Black Soldiers*, 66.
[422] Military Register for Dewitt C. Howell, Ancestry.com, Fold3.com.
[423] Company E Morning Reports; and Personnel Rosters, 1942 and 1943, NPRC.

[424] Company E Morning Reports, through 1942 and 1943, NPRC.
[425] Enlistment record of Robert William Lyons, Ancestry.com; Robert William Lyons obituary, Newspapers.com.
[426] H&S Company Personnel Rosters, officers, 1942 and 1943, NPRC.
[427] Company E Morning Reports, 1942–1943, NPRC.
[428] *General Court-Martial* CM-239042 testimony, 1943, NPRC.
[429] H&S Company Morning Reports, 1942–1943, NPRC.
[430] Twichell, *Northwest Epic*, 204.
[431] Regimental Progress Reports, 97th Regiment, 1942–1943, WP.
[432] Company F Morning Report March 10, 1943, NPRC.
[433] Intellicast.com, March 10, 1943.
[434] H&S Company Morning Report, March 12, 1943, NPRC.
[435] Wilkerson, *Warmth of Other Suns*, 31.

Notes to Chapter 19

[436] Brian Brettschneider, climatologist, International Arctic Research Center, Anchorage, Alaska, interview by author, 2019.
[437] The sources for this chapter regarding the events at Big Gerstle on March 29, 1943, include the following:

F and H&S Company Morning Reports, NPRC.

Cpt. Jack F. Doyle, investigating officer, Report of March 29, 1943, Incident; Sworn statements from accused: Pvt. James M. Heard, Pvt. Willie L. Howell, Pvt. Sims Bridges, Pvt. James V. Hollingsworth, Pvt. Josh Weaver, Pvt. Robert M. Rucker, Pvt. Willie B. Calhoun, Pvt. Lee I. Ratliff, Pvt. Eugene Fulks, and Pvt. Warren H. Lindsey; Sworn statements from witnesses: 1st Lt. DeWitt C. Howell, 2nd Lt. Robert W. Lyon Jr, 1st Sgt. Noah Williams, Tech 5 Willie L. Stallworth, Tech 5 Howard M. Graham, and Pfc. Clinton L. Hinson, April 1, 1943, all in NPRC.

General Court-Martial CM-239042 testimony, 1–450, NPRC.

Wagley, psychiatric reports of Willie Howell and James M. Heard at Fort Custer Rehabilitation Center, Michigan, NPRC.

Notes to Chapter 20

[438] H&S Company Morning Report, March 29,1943, NPRC.
[439] Lt. Col. J. W. Fraser, adjutant general for Brigadier General O'Connor, memorandum to Staff Judge Advocate, Northwest Service Command (NWSC), Whitehorse, Canada. Charge sheets. Approval of recommendations of Trial Judge Advocate. Recommendations for preparation of trial. Record of trial will be forwarded by endorsement to NWSC HQ. April 5 and April 7, 1943, NPRC.
[440] Chris Bray, *Court-Martial: How Military Justice Has Shaped America from the Revolution to 9/11 and Beyond* (New York: W. W. Norton, 2016), xiii.

441 Ibid., 265.
442 Rust, *Role of the Eighteenth Engineer Regiment*, 7.
443 Mitchim, Lt. Col. Charles F Formal Order Appointing Cpt. Jack Doyle to conduct an investigation into the charges against the ten accused mutineers, April 1, 1943, NPRC.
444 1st Sgt. Noah Williams Sworn Statement to Cpt. Jack F. Doyle during his investigation of charges against the ten accused. April 1, 1943, NPRC.
445 Cpt. Jack F. Doyle, investigating officer, Report on March 29 incident and recommendations; Lt. Col Charles F. Mitchim, Memorandum to Northwest Service Command forwarding Doyle's report and endorsing his recommendations, April 1943, NPRC.
446 Lt. Col. J. W. Fraser, Adjutant General for Brigadier General O'Connor, Memorandum to Staff Judge Advocate, Northwest Service Command. Whitehorse, Canada. Charge sheets. Approval of recommendations of Trial Judge Advocate. Recommendations for preparation of trial. Record of trial will be forwarded by endorsement to NWSC HQ, April 5 and April 7, NPRC. A lieutenant was assigned to determine whether the ten had any previous convictions and he duly reported that they had not.
447 Fraser, memo to Staff Judge Advocate, April 7, 1943, NPRC.
448 H&S Company Morning Report, April 5, 1943, NPRC.
449 Family biography, Ancestry.com; "To Be in Second Officer's Camp," *Star Press* (Muncie, Indiana), Sunday, August 12, 1917; "Dehorty Commissioned," *Call Leader* (Elwood, Indiana), Wednesday, November 28, 1917; "290 Students Receive Degrees from Indiana," *Indianapolis News*, Wednesday, June 11, 1919.
450 WWII Draft Card, Ancestry.com.
451 WWII Draft Card, Ancestry.com; Regimental Progress Reports, WP.
452 "J. Ward Starr Named to GOP Central Office" *Alexandria Times-Tribune*, Friday, May 23, 1952, Newspapers.com. The article contains reference to his election as prosecuting attorney and entering the army instead in 1942.
453 *General Court-Martial* CM-239042 testimony, June 5, 1943, through June 7, 1943, NPRC.
454 Letters, 1942–1943, WP.
455 Col. K. B. Bush, chief of staff, Headquarters, Northwest Service Command, Whitehorse, Yukon Territory, Canada, "Detail of Court," Special Order, May 29, 1943, NPRC.
456 Ibid., June 3.

Notes to Chapter 21
457 *A Manual for Courts Martial, U.S. Army*, Revised in the Office of the Judge Advocate General of the Army, corrected April 1940 (Washington, DC: US Government Printing Office 1943).
458 Ibid., 16.
459 Ibid., 27.
460 Ibid., 30.
461 Ibid., 231.
462 Company F Morning Reports, April 1, 1943, NPRC.

[463] Military Register, Ancestry.com.

[464] Ibid.; "Dr. Karl D. Reyer Joins LSU Staff," *The Times* (Shreveport, Louisiana), Thursday, June 24, 1937.

[465] Ancestry.com, 1940 Census, Military Register and Military Record, Beneficiary Identification Records Locator Subsystem (BIRLS) Records.

466 Bush, "Detail of Court," 1943, NPRC.

[467] *A Manual for Courts-Martial, U.S. Army* (Washington, DC: US Government Printing Office, 1928), 28.

[468] Col. K. B. Bush, Chief of Staff 1943, Headquarters Northwest Service Command, Whitehorse, Y. T Canada, "Detail of Court," Special Order No. 136, June 3, 1943, NPRC.

[469] Military Register, Military Record, US Army Directory, Ancestry.com; "Captain Whitney Quiets Mexican Border Trouble," *The Courier* (Waterloo, Iowa), Saturday, May 21, 1927; "Fort Bliss," *Daily Press* (Newport News, Virginia), Wednesday, July 17, 1929.

[470] Military Register, Ancestry.com; "War Department and Navy Orders," *Daily Press* (Newport News, Virginia), Saturday, April 10, 1937; "Harrison Barrack Space Increased with CC Quarters," *Vidette-Messenger of Porter Country* (Valparaiso, Indiana), Friday, August 14, 1942.

[471] Bray, *Court-Martial*, xiii.

[472] Walter Parsons III interview.

[473] Ibid.

[474] Ibid.

[475] Ibid.

[476] Ibid.

[477] Source for the remainder of the chapter is *General Court-Martial* CM-239042, 1–450.

Notes to Chapter 22

[478] Source for this chapter is *General Court-Martial* CM-239042, 1–450, June 5 through June 7, 1943, NPRC.

Notes to Chapter 23

[479] Source for this chapter is *General Court-Martial* CM-239042, 1–450, June 5 through June 7, 1943, NPRC.

Notes to Chapter 24

[480] Source for this chapter is *General Court-Martial* CM-239042, 1–450, June 5 through June 7, 1943, NPRC.

Notes to Chapter 25

[481] Source for this chapter is General Court-Martial CM-239042, 1–450, June 5 through June 7, 1943, NPRC.

[482] Letters, 1942–1943, WP.

Notes to Chapter 26

[483] Morning Reports, Station and Events, 97th Regiment, January 1942 to September 1943, NPRC.

[484] Company F Morning Report, Station and Events, April 1942 to September 1943, NPRC.

[485] Lt. Col. Samuel B. Lippitt, Staff Judge Advocate, "Review of the Staff Judge Advocate for Commanding Officer, Northwest Service Command," July 27, 1943, 1–10, NPRC; O'Connor, Brig. Gen. James A., Memorandum stating periods of and location of confinement and suspension of dishonorable discharge, July 30, 1943, NPRC; O'Connor, Brig Gen. James A., Memorandum stating findings and sentences disapproved for privates Eugene Fulks, Josh Weaver, James V. Hollingsworth, and Warren H. Lindy, July 30, 1943, NPRC.

[486] Ibid.
[487] Ibid.
[488] Ibid.
[489] Ibid.
[490] Ibid.

[491] Special Order 248, "Movement of Sixth Service Command Rehabilitation Center from Fort Custer, Michigan, to Fort Sheridan, Illinois, Service Unit 1662nd," October 16, 1943; Company Morning Report, Fort Custer, Michigan, 1641st Service Unit transferred to Fort Sheridan, Illinois, October 18, 1943, NPRC.

[492] Military Unit Locations and Dates, 1941–1944, 97th Engineer (GS) Regiment, First and Second 97th Battalion, 97th Headquarters and Service Company, 97th Company A-F; 97th Medical Detachment, Alaska, New Guinea, and Philippine Islands, June 1, 1941–April 24, 1944, NA.

[493] Ibid.

[494] "Army's Problem Children Are Making Good: Rehabilitation Plan Saves Delinquents," *Louisville Courier Journal* (Kentucky), June 4, 1943; "Rehabilitation Center Restores 40 Per Cent of Bad Actors to Good Soldiers," *Owensboro Messenger* (Kentucky), April 9, 1944.

[495] Robert B. Patterson, Undersecretary of War, Address at Conference on the Rehabilitation of Military Prisoners, November 15, Ft. Leavenworth, Kansas, RG 389, 1-3, NA.

[496] "Court martial Sentenced Men Get Second Opportunity at Bowie's Rehabilitation Center: 65 Percent Are Salvaged," Austin American-Statesman (Texas), August 1, 1943.

[497] Ibid.

[498] "Army's Problem Children Are Making Good: Rehabilitation Plan Saves Delinquents," *Louisville Courier Journal* (Kentucky), June 4, 1943.

[499] Ibid.
[500] Col. Alton C. Miller, Director, Provost Division, to Col. G. A. Gottschalk, Deputy Director, "Control of Mass or Group Movements in Correctional Institution," November 2, 1945, RG 389, NA.
[501] Ibid.
[502] "Army's Problem Children Are Making Good: Rehabilitation Plan Saves Delinquents," *Louisville Courier Journal* (Kentucky), June 4, 1943.
[503] "Court martial Sentenced Men Get Second Opportunity at Bowie's Rehabilitation Center," *Austin American-Statesman* (Texas), Sunday, August 1, 1943.
[504] Ibid.
[505] Austin H. MacCormick, Consultant to the Undersecretary of War, Address at Conference on the Rehabilitation of Military Prisoners, Fort Leavenworth, Kansas, November 16, 1944, RG 389, NA.
[506] Special Order 248, October 1943, NPRC.
[507] Military Unit Locations and Dates, 1941–1944, 97th Engineer (GS) Regiment, NA.
[508] Special Order 68, "Transferred enlisted, officers and prisoners from Sixth Service Command, Fort Sheridan Rehabilitation Center, Service Unit 1662nd, to Seventh Service Command Rehabilitation Center, Camp Philips, Kansas," March 20, 1944; Company Morning Report, Fort Sheridan, Illinois, 1662nd disbanded and personnel and prisoners transferred to Camp Phillips, Kansas, March 23, 1944, NPRC.
[509] Military Unit Locations and Dates, 1941–1944, 97th Engineer (GS) Regiment, NA.
[510] Brig. Gen. P. X. English, chief of staff, by command of Major General Danielson, General Court-Martial Order No. 652, Headquarters Seventh Service Command, restoring to duty Willie L. Howell, James M. Heard, Sims Bridges, Robert M. Rucker, Lee I. Ratliff, June 1, 1944, NPRC.
[511] Wilkerson, *Warmth of Other Suns*, 31.

Notes to Epilogue

[512] Willie B. Calhoun, US Army Service Record, December 1943, NPRC.
[513] Ibid., and US Army Medical Record, Incident Occurred August 1942, NPRC.
[514] Ibid., May 4, 1943, NPRC.
[515] Ibid., December 31,1943–January 4, 1944, NPRC.
[516] Ancestry.com.
[517] "Willie Ben Calhoun Shot," *Atlanta Constitution*, February 5, 1974.
[518] "101 Mississippians Dock in New York on John Ericsson," *Clarion-Ledger* (Jackson, Mississippi), Thursday, October 11, 1945.
[519] "Murder Charge Lodged against Vicksburg Negro," *Vicksburg Evening Post*, Wednesday, October 9, 1946.
[520] Ancestry.com, North Carolina Marriage Records; obituary, *Wilson Times*, February 7, 2001.
[521] Obituary, *Wilson Times*, February 7, 2001.

522 Josh (Joseph) Weaver, 1942–1943, US Army Service Record, Hospital Admission Card lists arm of service, 1945, and 97th Engineer Morning Reports, NPRC.
523 Obituary, *Savannah News Press*, "Joseph Weaver," February 15, 1987, Chatham County, Georgia.
524 James Hollingsworth, 1942–1943, US Army Service Record, Final Payment Record, January 10, 1945, NPRC; Ancestry.com, Georgia Death Index 1919–1998.
525 US Army Service Record, Final Payment Record, January 25, 1946, NPRC. And Ancestry.com, Find a Grave headstone.
526 Diana Rucker Phifer, interview by author, 2018.
527 Ibid.
528 Ibid.
529 Ibid.
530 Ancestry.com, BIRLS Death File, and Find a Grave.
531 "Howell, "Obituary" *Atlanta Constitution*, Friday, August 1, 2001.
532 US Army Service Record, Final Payment Record, January 5,1946, NPRC.
533 Les Young, interview by author, 2018.
534 Ibid.
535 Ibid.
536 Ancestry.com, Death Record.
537 US Army Service Records, General Court-Martial Orders No. 515, May 12, 1944, NPRC; Ancestry.com, BIRLS Death File.
538 Ancestry.com.
539 Ancestry.com, BIRLS Death File.
540 US Army Service Records, Court-Martial Record, returned to duty May 14, 1944, suspension vacated April 19, 1945 (assaulting a woman in France), February 14, 1946, a War Department Memo, July 16, 1946, confined at Penitentiary, Terre Haute, Indiana, NPRC.
541 "Prisoners Will Be Arraigned in Ontario County Court Tomorrow," *Daily Messenger* (Canandaigua, New York), October 21, 1952.
542 "Theft Suspect Held," *Democrat and Chronicle* (Rochester, New York), Tuesday, February 4, 1958.
543 "Sims Bridges," *Democrat and Chronicle* (Rochester, New York), Tuesday, November 19, 1963.
544 E. D. Johnson, New York State Department of Correctional Services (COCCS) at Attica, October 28, 2019.
545 Ibid.

SELECT BIBLIOGRAPHY

Archival Sources

CB Camp Blanding Museum, Starke, Florida
NA National Archives, College Park, Maryland
NPRC National Personnel Records Center, St. Louis, Missouri
UAF University of Alaska, Fairbanks, Alaska, Elmer E. Rasmuson Library, Alaska and Polar Regions Collections
USACE US Army Corps of Engineers Office of History, Humphreys Engineer Center, Fort Belvoir, Virginia
USAHEC US Army Heritage and Education Center, Carlisle, Pennsylvania
WP Walter Parsons private collection

Books

Abercrombie, Cpt. W. R. Alaska, 1899, *Copper River Exploring Expedition*. Washington, DC: US Government Printing Office, 1900.

Blasor-Bernhardt, Donna. *Pioneer Road: Recollections of the Pioneers Who Built the Alaska Highway*. Las Vegas: ArcheBooks Publishing, 2004.

Bleakley, Geoffrey. *History of the Valdez Trail*. Wrangell-St. Elias National Park and Preserve, Alaska, April 14, 2015.

Boyd, Robert Platt. *Me and Company "C."* N.p.: Self-published, 1992.

Bray, Chris. *Court-Martial: How Military Justice Has Shaped America from the Revolution to 9/11 and Beyond*. New York: W. W. Norton, 2016.

Butler, Susan. *Roosevelt and Stalin: Portrait of a Partnership*. New York: Vintage Books, 2015.

Camp, Lynn Robinson. *Morgan County, Georgia. Black America Series*. Charleston, SC: Arcadia Publishing, 2004.

Caterpillar Tractor Company, *Fifty Years on Tracks*. Chicago: Photopress, 1954.

Cohen, Stan. *ALCAN and CANOL: A Pictorial History of Two Great World War II Construction Projects*. Missoula, MO: Pictorial Histories Publishing, 1992.

Cohen, Stan. *The Forgotten War: A Pictorial History of World War II in Alaska and Northwestern Canada.* Madison, WI: Pictorial Histories Publishing, 1981.

Coll, Blanche D., Jean E. Keith, and Herbert H. Rosenthal. *United States Army in World War II: The Corps of Engineers: Troops and Equipment.* Washington, DC: US Government Printing Office, 1958.

Dorsey, Albert, Jr. "A Mississippi Burning: Examining the Lynching of Lloyd Clay and the Encumbering of Black Progress in Mississippi during the Progressive Area." Unpublished manuscript, Spring 2009. http://diginole.lib.fsu.edu/islandora/object/fsu:168847/datastream/PDF/view.

Duesenberg, H. Milton. *Alaska Highway Expeditionary Force: A Roadbuilder's Story.* Clear Lake, IA: H & M Industries Ltd., 1995.

Garfield, Brian. *The Thousand Mile War: World War II in Alaska and the Aleutians.* Fairbanks: University of Alaska Press, 1969.

Godsell, Philip H. *Romance of the Alaska Highway.* Toronto: Ryerson Press, 1944.

Griggs, William E. *The World War II Black Regiment That Built the Alaska Military Highway* ed. Philip J. Merrill. Jackson: University Press of Mississippi, 2002.

Hamm, Aurolyn Melba. *Elbert County, Georgia, Black American Series.* Charleston, SC: Arcadia Publishing, 2005.

Harrais, Margaret Keenan. *Remembering Old Valdez.* Unpublished Autobiography, n.d. Archived at Alaska Room, Valdez Alaska Library.

Hegener, Helen. *Alaskan Roadhouses: Shelter, Meals, and Lodging along Alaska's Early Roads and Trails.* Wasilla, AK: Northern Light Media Publishers, 2015.

Kane, Sharyn, and Richard Keeton. *In Those Days: African American Life Near the Savannah River.* Atlanta: National Park Service, Southeast Region, 1994.

Lee, Ulysses. *The Employment of Negro Troops.* Washington, DC: US Government Printing Office, 1963.

Lincoln, C. Eric, and Lawrence H. Mamiya. *The Black Church in the African American Experience.* Durham, NC: Duke University Press, 1990.

Margo, Robert A. *Race and Schooling in the South: A Review of the Evidence.* Chicago: University of Chicago Press, 1990.

Marsh, Kenneth L. *The Trail: The Story of the Historic Valdez–Fairbanks Trail That Opened Alaska's Vast Interior.* Trapper Creek, AK: Trapper Creek Museum, 2008.

McKinney, Charles W., Jr. *Greater Freedom: The Evolution of the Civil Rights Struggle in Wilson, North Carolina.* Lanham, MD: University Press of America, 2010.

Medical Department, Department of the Army, *Preventive Medicine in World War II,* vol. 5: *Communicable Diseases Transmitted through Contact or by Unknown Means,* ed. Lt. Gen. Leonard D. Heaton and Col. John Boyd Coates Jr. Washington, DC: Office of the Surgeon General, 1960.

Naske, Claus M. *Paving Alaska Trails: The Work of the Alaska Road Commission.* Lanham, MD: University Press of America, 1986.
Schmidt, John. *This Was No DYXNH Picnic.* Alberta, Canada: Gorman and Gorman, 1991.
Tower, Elizabeth A. *Icebound Empire: Industry and Politics on the Last Frontier, 1898–1938*, 3rd ed. Louisville, KY: Old Stone Press, 2015.
Twichell, Heath. *Northwest Epic: The Building of the Alaska Highway.* New York: St. Martin's Press, 1992.
US War Department. *A Manual for Courts-Martial, U.S. Army.* Revised in the Office of the Judge Advocate General of the Army. Corrected April 1940. Washington, DC: US Government Printing Office, 1936. Originally published 1928.
Virtue, John. *The Black Soldiers Who Built the Alaska Highway: A History of Four U.S. Army Regiments in the North, 1942–1943.* Jefferson, NC: McFarland, 2013.
Wilkerson, Isabel. *The Warmth of Other Suns.* New York: Vintage Books, 2010.

ABOUT THE AUTHORS

Christine and Dennis McClure married in 1992. Christine was born in Annapolis, MD, and later served in the United States Army as a registered nurse at the end of the Vietnam era. Dennis was born and raised in Northern Michigan, pursued a PhD in history at Cornell University and served in the United States Army. The couple lives in Taylors, South Carolina.

www.ingramcontent.com/pod-product-compliance
Lightning Source LLC
Chambersburg PA
CBHW020904080526
44589CB00011B/431